T0257633

# Encyclopedia of Alternative and Renewable Energy: AC Offshore Wind Farms

## Volume 19

# Encyclopedia of Alternative and Renewable Energy: AC Offshore Wind Farms Volume 19

Edited by **Benjamin Wayne and David McCartney**

New York

Published by Callisto Reference,
106 Park Avenue, Suite 200,
New York, NY 10016, USA
www.callistoreference.com

**Encyclopedia of Alternative and Renewable Energy:**
**AC Offshore Wind Farms**
**Volume 19**
Edited by Benjamin Wayne and David McCartney

© 2015 Callisto Reference

International Standard Book Number: 978-1-63239-193-3 (Hardback)

Printed in the United States of America.

# Contents

**Permissions**

# Preface

Over the recent decade, advancements and applications have progressed exponentially. This has led to the increased interest in this field and projects are being conducted to enhance knowledge. The main objective of this book is to present some of the critical challenges and provide insights into possible solutions. This book will answer the varied questions that arise in the field and also provide an increased scope for furthering studies.

The book aims at providing valuable information to the readers regarding AC offshore wind farms. It covers vital aspects of offshore wind farm's energy transmission and grid integration infrastructure. However, not all the electric configurations have been evaluated for this purpose. This book presents a representative case which has been comprehensively analyzed. It has been constructed on three general features of an offshore wind farm: the rated power, the distance to shore and the average wind speed of the place. Thus, after a concise account of concepts related to wind power and several subsea cable modeling options, an offshore wind farm has been modeled and its parameters have been defined to use as a case study. Upon this case study, several analyses of key aspects of connection infrastructure have been performed. The primary aspect to be determined was the management of reactive power flowing through the submarine cable. The much unwelcomed harmonic amplifications in offshore wind farms due to resonances and after this, transient over-voltage complications in electric framework have been characterized. Finally, an offshore wind farm connection infrastructure has been presented in order to meet the grid code requirements for a specific system operator, but not as a close solution, as a result of a methodology based on analysis and simulations to define the most appropriate layout depending on the size and location of each offshore wind farm.

I hope that this book, with its visionary approach, will be a valuable addition and will promote interest among readers.

**Editor**

# Abstract

The best places to build a wind farm in land are in use, due to the spectacular growth of the wind power over the last decade. In this scenario offshore wind energy is a promising application of wind power, particularly in countries with high population density, and difficulties in finding suitable sites on land.

On land wind farms have well-adjusted their features and the transmission system to each wind farm size and characteristic. But for offshore wind farms this is an open discussion.

This book analyses the offshore wind farm's electric connection infrastructure, thereby contributing to this open discussion. So, a methodology has been developed to select the proper layout for an offshore wind farm for each case. Subsequently a pre-design of the transmission system's support equipment is developed to fulfill the grid code requirements.

# Chapter 1

# Introduction

Wind energy is one of the most important energy resources on earth. It is generated by the unequal heat of the planet surface by the sun. In fact, 2 per cent of the energy coming from the sun is converted into wind energy. That is about 50 to 100 times more than the energy converted into biomass by plants.

Several scientific analyses have proven wind energy as a huge and well distributed resource throughout the five continents. In this way, the European Environment Agency in one of its technical reports evaluating the European wind potential [1], estimates that this potential will reach 70.000TWh by 2020 and 75.000TWh by 2030, out of which 12.200TWh will be economically competitive potential by 2020. This amount of energy is enough to supply three times the electricity consumption predicted for this year (2020). The same study also evaluates the scenario in 2030 when the economically competitive potential increases to 200TWh, seven times the electricity consumption predicted for this year (2030).

Today electricity from wind provides a substantial share of total electricity production in only a handful of Member States (see Figure 1.1), but its importance is increasing. One of the reasons for this increment is the reliability of this energy resource, which has been proven from the experience in Denmark. In this country 24% of the total energy production in 2010 was wind-based and the Danish government has planned to increase this percentage to 50% by 2030.

Following Denmark, the countries with the highest penetration of wind power in electricity consumption are: Portugal (14.8%), Spain (14.4%) and Ireland (10.1%)

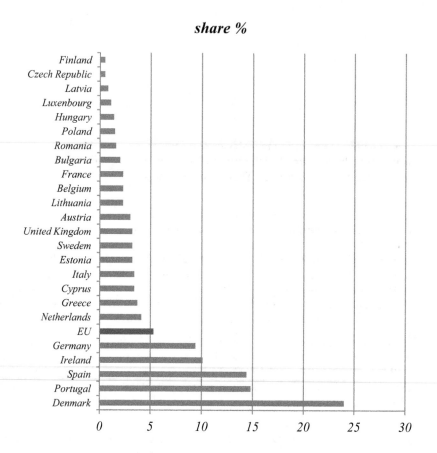

**share %**

Figure 1.1 Wind share of total electricity consumption in 2010 by country [2].

This spectacular growth of the wind power share in the electricity consumption is supported in the new installed wind power capacity. In this way, more than 40% of all new electricity generation capacity added to the European grid in 2007 was wind-based [4]. However, this year was not an exception, wind power is been the fastest growing generation technology except for natural gas in the decade (2000-2010), see Figure 1.2.

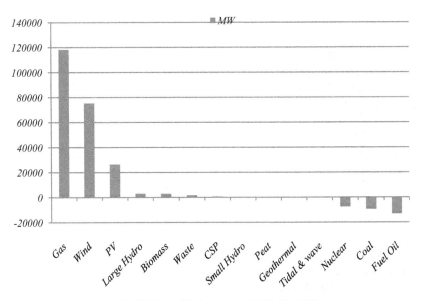

Figure 1.2 Net changes in the EU installed capacity 2000-2010 [2].

The considered scenario used by the European Union for the Second Strategic Energy Review [5] suggests that wind will represent more than one third of all electricity production from renewable energy sources by 2020 and almost 40% by 2030, representing an accumulated investment of at least 200-300 billion Euros (or about a quarter of all power plant investments) by 2030.

Due to the fast growth of the onshore wind energy exposed before, in many countries the best places to build a wind farm onshore are already in use, so in the future of this technology, offshore wind power is destined to have an important role. Because, offshore wind energy can be the way to meet the objectives of the new Energy Policy for Europe since it's an indigenous resource for electricity production, as well as clean and renewable.

Offshore wind can and must make a substantial contribution to meeting all three key objectives of EU's energy policy: Reducing greenhouse gas emissions, ensuring safety of supply and improving EU competitiveness in a sector in which European businesses are global leaders.

Nowadays, offshore wind energy is emerging and installation offshore wind farms at sea will become increasingly important. 430 MW of offshore wind power capacity were installed in 2009, the 4% of all the installed wind energy capacity. But, with 1107 MW of new installed capacity, 2010 was a record-breaking year for offshore wind power.

This trend is not only an issue of the last two years, offshore capacity has been gradually increasing since 2005 and in 2010 it represents around the 10% of all new wind power installations, see Figure 1.3.

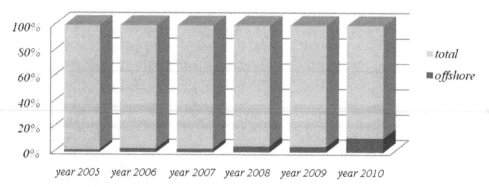

Figure 1.3 Offshore wind power share of total installed wind power capacity [2].

Furthermore, this energy resource will cover a huge share of the electricity demand, since the exploitable potential by 2020 is likely to be some 30-40 times the installed capacity in 2010 (2.94 GW) , and in the 2030 time horizon it could be up to 150 GW (see Figure 1.4), or some 575 TWh [5].

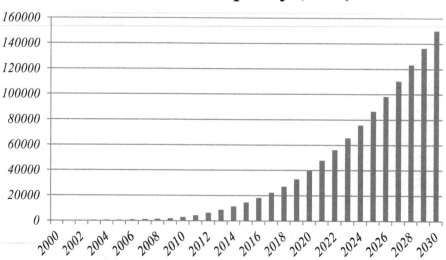

Figure 1.4 Estimation for offshore wind power capacity evolution 2000-2030 [3].

Wind energy is now firmly established as a mature technology for electricity generation and an indigenous resource for electricity production with a vast potential that remains largely untapped, especially offshore.

Thus, the EU is pushing a stable and favorable framework to promote offshore wind farms and renewable energy in general. To this end, it is implementing plans such as the third internal energy market package of October 2007 [6] or the energy and climate package presented in January 2008 [7].

Supported in this favorable framework, Europe has become the world leader in offshore wind power, especially United Kingdom and Denmark. The first offshore wind farm was being installed in Denmark in 1991 and in 2010 the United Kingdom has by far the largest capacity of offshore wind farms with 1.3 GW, around 40% of the world total capacity.

As regards of the rest of the countries of the union, only nine countries have offshore wind farms and most of them located in the North Sea, Irish sea and Baltic sea, Table 1.1.

| Country | Cumulative capacity (MW) | Installed capacity 2010 (MW) |
|---|---|---|
| Belgium | 195 | 165 |
| Denmark | 853.7 | 207 |
| Finland | 26.3 | 2.3 |
| Germany | 92 | 80 |
| Ireland | 25.2 | - |
| Netherlands | 246.8 | - |
| Norway | 2.3 | - |
| Sweden | 163.7 | - |
| United kingdom | 1341.2 | 652.8 |
| TOTAL | 2946.2 | 1107.1 |

Table 1.1 Offshore wind cumulative and installed capacity in 2010 by country

Nevertheless, offshore wind is not only an issue of the mentioned three seas in the European Union. In the south for example, Italy has planned around 4199.6 MW distributed in 11 wind farm projects for the upcoming years. French republic has also planned 3443,5 MW and three additional projects in the Mediterranean sea.

In the same way, the Iberian Peninsula is no exception to the growth and development of offshore energy. Offshore wind farms with 4466 MW total rated power are planned for the upcoming years, This means that the Iberian Peninsula has planned four times the offshore power in Europe in 2008. Even Croatia (392 MW) and Albania (539 MW) have planned offshore wind farms [8].

Furthermore, in the south/center of the European Union, there are two wind farms under construction one in Italy (90 MW, Tricase) and another one in France (105 MW, cote d'Albatre) located in the English channel.

As a result of these efforts, EU companies are leading the development of this technology in the world: Siemens and Vestas are the leading turbine suppliers for offshore wind power and DONG Energy, Vattenfall and E.ON are the leading offshore operators.

This evolution of the wind farms from onshore to offshore have led to some technological challenges, such as the energy transmission system or energy integration in the main grid.

Onshore wind farms have adjusted their characteristics well to the size and features of each wind farm as a result of the huge experience in this field. But for offshore, there are only a few built wind farm examples and the energy transmission is through submarine cables, so the definition of the most suitable layout is still an open discussion.

Offshore wind farms must be provided with reliable and efficient electrical connection and transmission system, in order to fulfill the grid code requirements. Nowadays, there are many and very different alternatives for the offshore wind farms transmission system configurations.

This is because the main difference in the transmission system between onshore wind farms and offshore wind farms is the cable used. Offshore wind farms need submarine cables. That present a high shunt capacitance in comparison to overhead lines [9]. The capacitive charging currents increase the overall current of the cable and thus reduce the power transfer capability of the cable (which is thermally limited).

Due to the spectacular growth of wind energy, many countries have modified their grid codes for wind farms or wind turbines requiring more capabilities. Some countries have specific grid codes referring to wind turbine/farm connections, such as Denmark, Germany or Ireland. The great majority of these countries have their grid code requirements oriented towards three key aspects: Power quality, reactive power control and Low Voltage Ride Through (LVRT).

The new grid code requirements are pushing new propositions in fields like power control, power filters or reactive power compensation, with new control strategies and components for the transmission system in order to integrate energy into the main grid.

These propositions have strong variations depending on the grid codes and the different kind of transmission systems such as: Medium Voltage Alter Current (MVAC) configurations or High Voltage Direct Current (HVDC) configurations.

For onshore wind farms, depending on the size and location features, their characteristics are well adjusted. However, for offshore wind farms the definition of the most suitable layout is still an open discussion.

The objective of this book is to contribute to this open discussion analyzing the key issues of the offshore wind farm's energy transmission and grid integration infrastructure. But, for this purpose, the objective is not the evaluation of all the electric configurations. The aim of the present book is to evaluate a representative case.

The definition of the electric connection infrastructure, starting from three generic characteristics of an offshore wind farm: the rated power of the wind farm, the distance to shore and the average wind speed of the location. In this way, it is possible to identify the problematic aspects of the energy transmission and grid integration based on this representative specific case.

In short, the development of an evaluation and simulation methodology to define the most suitable layout depending on the size and location of each wind farm, as for the onshore wind farms. This pre-design has to be suitable to connect to a distribution grid. Therefore, it has to fulfill the grid code requirements.

To accomplish this goal, this book contributes to the better knowledge of the nature, the causes and the problematic aspects of the electric connection infrastructure. The following key issues are evaluated.

- Submarine cable modeling options and the accuracy of those models.
- The influence of the main components of the offshore wind farm in its frequency response is analyzed, to help avoiding harmonic problems in the offshore wind farm at the pre-design stage.
- Transient over-voltage problems in the electric infrastructure of the offshore wind farms are characterized, more specifically, transient over-voltages caused by switching actions and voltage dips at the PCC.

Then, based on those evaluations of the key issues of the electric connection infrastructure, several solutions to fulfill the grid codes are proposed and tested via simulation:

- The management of the reactive power through the submarine power cable is evaluated and dimensioned for a specific case.
- The passive filters are dimensioned for the considered specific case. Furthermore, the most suitable location for these filters is analyzed (onshore / offshore).
- The auxiliary equipment to protect the offshore wind farm upon switching actions and fault clearances are discussed.
- The auxiliary equipment to fulfill the grid codes during voltage dips at the PCC are dimensioned.

# Chapter 2

# Wind Energy

The aim of this chapter is to introduce the reader to the wind energy. In this way, as the primary source of wind energy, how the wind is created and its characteristics are evaluated.

Due to its nature, the wind is an un-programmable energy source. However, it is possible to estimate the wind speed and direction for a specific location using wind patterns. Therefore, in the present chapter, how to describe the wind behavior for a specific location, the kinetic energy contained in the wind and its probability to occur is described.

To convert the wind energy into a useful energy has to be harvested. The uptake of wind energy in all the wind machines is achieved through the action of wind on the blades, is in these blades where the kinetic energy contained in the wind is converted into mechanic energy. Thus, the different ways to harvest this energy are evaluated, such as: different kind of blades, generators, turbines...

Once, the wind and the fundamentals of the wind machines are familiar, the advantages / disadvantages between offshore and onshore energy are discussed.

## 2.1 The wind
The unequal heat of the Earth surface by the sun is the main reason in the generation of the wind. So, wind energy is a converted form of solar energy.

The sun's radiation heats different parts of the earth at different rates; this causes the unequal heat of the atmosphere. Hot air rises, reducing the atmospheric pressure at the earth's surface, and cooler air is drawn in to replace it, causing wind. But not all air mass displacement can be denominate as wind, only horizontal air movements. When air mass has vertical displacement is called as "convection air current"

The wind in a specific location is determinate by global and local factors. Global winds are caused by global factors and upon this large scale wind systems are always superimposed local winds.

Global or geostrophic winds

The geostrophic wind is found at altitudes above 1000 m from ground level and it's not very much influenced by the surface of the earth.

The regions around equator, at 0° latitude are heated more by the sun than the regions in the poles. So, the wind rises from the equator and moves north and south in the higher layers of the atmosphere. At the Poles, due to the cooling of the air, the air mass sinks down, and returns to the equator.

If the globe did not rotate, the air would simply arrive at the North Pole and the South Pole, sink down, and return to the equator. Thus, the rotation with the unequal heating of the surface determines the prevailing wind directions on earth. The general wind pattern of the main regions on earth is depicted in Figure 2.1

Figure 2.1 Representation of the global wind on the earth.

Besides the earth rotation, the relative position of the earth with the sun also varies during the year (year seasons). Due to these seasonal variations of the sun's radiation the intensity and direction of the global winds have variations too.

Local Winds

The wind intensity and direction is influenced by global and local effects. Nevertheless, when global scale winds are light, local winds may dominate the wind patterns. The main local wind structures are sea breezes and mountain / valley breezes. The breeze is a light and periodic wind which appears in locations with periodic thermal gradient variations.

Figure 2.2 Illustration of the sea breezes direction.

Figure 2.3 Illustration of the mountain / valley breezes direction.

Land masses are heated by the sun more quickly than the sea in the daytime. The hot air rises, flows out to the sea, and creates a low pressure at ground level which attracts the cool air from the sea. This is called a sea breeze. At nightfall land and sea temperatures are equal and wind blows in the opposite direction [10].

A similar phenomenon occurs in mountain / valleys. During the day, the sun heats up the slopes and the neighboring air. This causes it to rise, causing a warm, up-slope wind. At night the wind direction is reversed, and turns into a down-slope wind.

### 2.1.1 The roughness of the wind

About 1 Km above the ground level the wind is hardly influenced by the surface of the earth at all. But in the lower layers of the atmosphere, wind speeds are affected by the friction against the surface of the earth. Therefore, close to the surface the wind speed and wind turbulences are high influenced by the roughness of the area.

In general, the more pronounced the roughness of the earth's surface, the more the wind will be slowed down.

Trees and high buildings slow the wind down considerably, while completely open terrain will only slow the wind down a little. Water surfaces are even smoother than completely open terrain, and will have even less influence on the wind.

The fact that the wind profile is twisted towards a lower speed as we move closer to ground level is usually called wind shear. The wind speed variation depending on the height can be described with the following equation (1) [11]:

$$\frac{V'_{wind}}{V_{wind}} = \left(\frac{h'}{h}\right)^{\alpha_w} \tag{1}$$

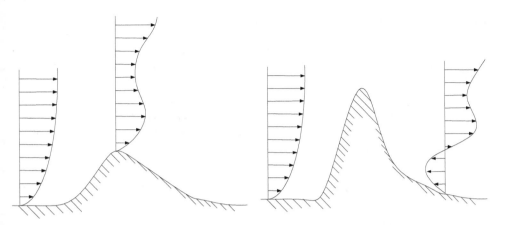

Figure 2.4 Illustration of the wind speed variation due to the obstacles in the earth surface.

Where: $V'_{wind}$= the velocity of the wind (m/s) at height $h'$ above ground level. $V_{wind}$ = reference wind speed, i.e. a wind speed is already known at height $h$. $h'$ = height above ground level for the desired velocity, $\alpha_w$ = roughness length in the current wind direction. $h$ = reference height (the height where is known the exact wind speed, usually =10m).

As well as the wind speed the energy content in the wind changes with the height. Consequently, the wind power variations are described in equation (2) [12]:

$$\frac{P'_{wind}}{P_{wind}} = \left(\frac{h'}{h}\right)^{3\alpha_w} \tag{2}$$

Where: $P'_{wind}$ = wind power at height $h'$ above ground level. $P_{wind}$ = reference wind power, i.e. a wind power is already known at height $h$. $h'$ = height above ground level for the desired velocity, $\alpha_w$ = roughness length in the current wind direction. $h$ = reference height (the height where is known the exact wind speed, usually =10m).

At the following table, the different values of $\alpha_w$ (The roughness coefficient) for different kind of surfaces, according to European Wind Atlas [13] are shown.

| 0 0,0002 | Water surface |
|---|---|
| 0,0024-0,5 | Completely open terrain with a smooth surface, e.g. concrete runways in airports, mowed grass, etc. |
| 0,03-1 | Open agricultural area without fences and hedgerows and very scattered buildings. Only softly rounded hills |
| 0,4-3 | Villages, small towns, agricultural land with many or tall sheltering hedgerows, forests and very rough and uneven terrain |
| 1,6-4 | Very large cities with tall buildings and skyscrapers |

Table 2.1 Different α values for different kind of surfaces.

## 2.1.2 The general pattern of wind: Speed variations and average wind

Wind is an un-programmable energy source, but this does not mean unpredictable. It is possible to estimate the wind speed and direction for a specific location. In fact, wind predictions and wind patterns help turbine designers to optimize their designs and investors to estimate their incomes from electricity generation.

The wind variation for a typical location is usually described using the so-called "Weibull" distribution. Due to the fact that this distribution has been experimentally verified as a pretty accurate estimation for wind speed [14], [15] The weibull's expression for probability density (3) depends on two adjustable parameters.

$$\phi(v) = \frac{k}{c} \cdot \left( \frac{v_{wind}}{c} \right)^{k-1} \cdot e^{-\left( \frac{v_{wind}}{c} \right)^{k}} \tag{3}$$

Where: $\phi(v)$= Weibull's expression for probability density depending on the wind, $v_{wind}$ = the velocity of the wind measured in m/s, $c$ = scale factor and $k$ = shape parameter.

The curves for weibulls distribution for different average wind speeds are shown in Figure 2.5. This particular figure has a mean wind speed of 5 to 10 meters per second, and the shape of the curve is determined by a so called shape parameter of 2.

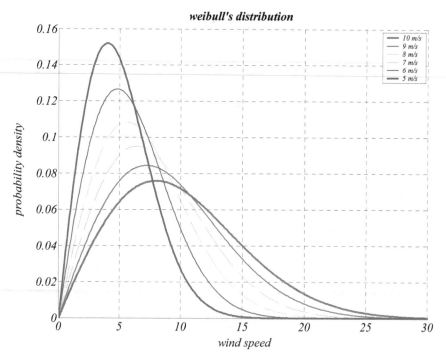

Figure 2.5 Curves of weibull's distribution for different average wind speeds 5, 6, 7, 8, 9 and 10 m/s.

The graph shows a probability density distribution. Therefore, the area under the curve is always exactly 1, since the probability that the wind will be blowing at some wind speed including zero must be 100 per cent.

The statistical distribution of wind speed varies from one location to another depending on local conditions like the surfaces roughness. Thus to fit the Weibull distribution to a specific location is necessary to set two parameters: the shape and the wind speeds mean value.

If the shape parameter is 2, as in Figure 2.5, the distribution is known as a Rayleigh distribution. Wind turbine manufacturers often give standard performance figures for their machines using the Rayleigh distribution.

The distribution of wind speeds is skewed, is not symmetrical. Sometimes the wind presents very high wind speeds, but they are very rare. On the contrary, the probability of the wind to presents slow wind speeds is pretty high.

To calculate the mean wind speed, the wind speed value and its probability is used. Thus, the mean or average wind speed is the average of all the wind speeds measured in this location. The average wind speed is given by equation (4) [16]:

$$\overline{v}_{wind} = \int_{0}^{\infty} v_{wind} \cdot \phi(v_{wind}) \cdot dv_{wind} \qquad (4)$$

Where: $\phi(v_{wind})$ = Weibull's expression for probability density depending on the wind, $v_{wind}$ =the velocity of the wind measured in m/s.

## 2.2 The power of the wind
The uptake of wind energy in all the wind machines is achieved through the action of wind on the blades, is in these blades where the kinetic energy contained in the wind is converted into mechanic energy. Therefore, in the present section the analysis of the power contained in the wind is oriented to those devices.

### 2.2.1 The kinetic energy of the wind
The input power of a wind turbine is through its blades, converting wind power into a torque. Consequently, the input power depends on the rotor swept area, the air density and the wind speed.

Air density

The kinetic energy of a moving body is proportional to its mass. So, the kinetic energy of the wind depends on the air density, the air mass per unit of volume. At normal atmospheric pressure (and at 15° C) air weigh is 1.225 kg per cubic meter, but the density decreases slightly with increasing humidity.

Also, the air is denser when it is cold than when it is warm. At high altitudes, (in mountains) the air pressure is lower, and the air is less dense.

## The rotor area

The rotor area determines how much energy a wind turbine is able to harvest from the wind. Due to the fact that the amount of the air mass flow upon which the rotor can actuate is determined by this area, this amount increases with the square of the rotor diameter, equation (5)

$$A_r = \pi \cdot r^2 \tag{5}$$

Where: $A_r$ = the rotor swept area in square meters and $r$ = the radius of the rotor measured in meters.

## Equation of the winds kinetic energy

The input air mass flow of a wind turbine with a specific rotor swept area determined by $A_r$ is given by equation (6). This input air mass flow depends on the wind speed and the rotor swept area.

$$M = \rho A_r v_{wind} \tag{6}$$

Where: $M$ = Air mass flow, $\rho$ = the density of dry air ( 1.225 measured in kg/m³ at average atmospheric pressure at sea level at 15° C) and $V_{wind}$ = the velocity of the wind measured in m/s.

Therefore, the winds kinetic energy is given by equation (7).

$$P_{wind} = \frac{1}{2} M v^2_{wind} = \frac{1}{2} \rho A_r v^3_{wind} \tag{7}$$

Where: $P_{wind}$ = the power of the wind measured in Watts, $\rho$ = the density of dry air ( 1.225 measured in kg/m³ at average atmospheric pressure at sea level at 15° C), $V_{wind}$ = the velocity of the wind measured in m/s and $r$ = the radius of the rotor measured in meters.

The wind speed determines the amount of energy that a wind turbine can convert to electricity. The potential energy per second in the wind varies in proportion to the cube of the wind speed, and in proportion to the density of the air.

### 2.2.2 Usable input power, Betz law

The more kinetic energy a wind turbine pulls out of the wind, the more the wind will be slowed down. In one hand if the wind turbines extract all the energy from the wind, the air could not leave the turbine and the turbine would not extract any energy at all. On the other hand, if wind could pass though the turbine without being hindered at all. The turbine would not extract any energy from the wind.

Therefore is possible to assume that there must be some way of breaking the wind between these two extremes, to extract useful mechanical energy from the wind.

## Betz law

Betz law says that it's only possible convert less than 16/27 (or 59%) of the kinetic energy in the wind to mechanical energy using a wind turbine. This law can be applied to any kind of wind generators with disc turbines. Besides this limit, also must be considered the aerodynamic and mechanic efficiency from the turbines.

### 2.2.3 Useful electric energy from wind

As said before, from the winds kinetic energy it's only possible convert less than 16/27 (Betz's law). However, the process to harvest the wind also has other losses, even the best blades have above 10% of aerodynamic losses [17].

So, the electric power that can be extracted from the kinetic energy of the wind with a turbine is given by the well-known equation (8).

$$P_t(v) = \frac{1}{2} \cdot \rho \cdot \pi \cdot r^2 \cdot V_{wind}^3 \cdot Cp \tag{8}$$

Where: $P_t(v)$ = the input power of the generator, $\rho$ = the density of dry air ( 1.225 measured in kg/m$^3$ for average atmospheric pressure at sea level with 15° C), $r$ = the radius of the rotor measured in meters, $V_{wind}$ = the velocity of the wind measured in m/s and $Cp$ = the power coefficient.

As any machine in movement, the generator has mechanic losses whether they are: at the bearings, brushes, gear...Equally any electric machine has electric losses. Hence, only a part of the winds kinetic energy can be converted to electric power Figure 2.6.

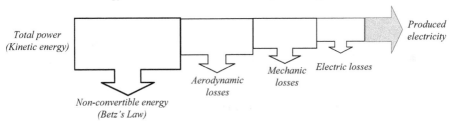

Figure 2.6 Representation of the power losses at different steps of electric wind energy generation.

### 2.2.3.1 The power coefficient

The power coefficient tells how efficient is a turbine capturing the energy contained in the wind. To measure this efficiency, the energy captured by the rotor is divided by the input wind energy. In other words, the power coefficient is the relation between the kinetic energy in the rotor swept area and the input power of the generator.

## 2.3 Fundamentals of wind machines

Wind machines convert the kinetic energy contained in the wind into mechanic energy through the action of wind on the blades. The aerodynamic principle in this transformation (kinetic to mechanic energy) is similar to the principle that makes airplanes fly.

According to this principle, the air is forced to flow over the top and bottom of a blade (see Figure 2.7) generating a pressure difference between both sides. The pressure difference causes a resultant force upon the blade. This force can be decomposed in two components:

a) *Lift force*, which is perpendicular to the direction of the wind.

b) *Drag force*, which is parallel to the direction of the wind. This force helps the circulation of air over the surface of the blades.

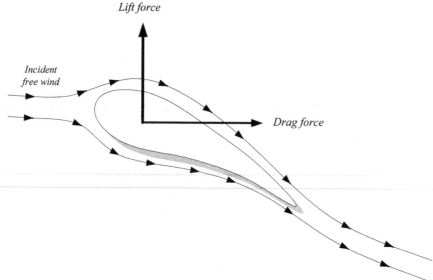

Figure 2.7 Representation of lift force and drag force generated on the blades.

The force which will generate a torque is lift force or drag force depending on the relative position of the blades with the axis and the wind.

In wind turbines with horizontal axis, the lift component of the force is the only one that gives the torque. Therefore, as the lift force gives torque, the profile of the blade has to be designed setting the attack angle ($\alpha$), the relative position of the blade with the wind (see Figure 2.8), to make maximum lift / drag force ratio [12].

This simple analysis is only valid when the blades of a wind turbine are at rest. If the rotation of the rotor is allowed, the resultant force on the blades will be the result from the combination of direct action of the real wind and the action of the wind created by the blades.

The incident wind on the blades is called apparent wind (Figure 2.8), is the result from the composition of the vector of the true wind vector and the wind created by the blade.

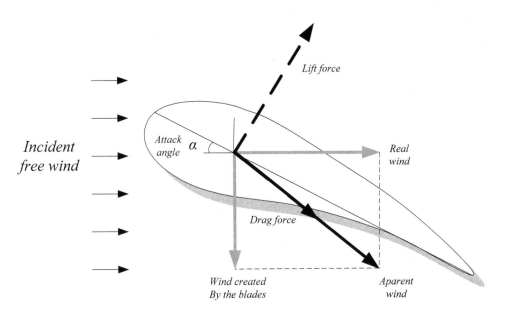

Figure 2.8 Wind created by the blade and the apparent wind.

Each section of the blade has a different speed and the wind speed is higher in terms of the height, thus, the apparent wind in each section is different. To obtain the same resultant force along its length, the profile of the blade has to have different dimensions. Therefore, to achieve this homogeneous resultant force, the rotor blade is twisted. The wing does not change its shape, but changes the angle of the wing in relation to the general direction of the airflow (also known as the angle of attack).

To start a wind turbine, wind speed must exceed the so-called cut in speed (minimum value needed to overcome friction and start producing energy) usually between 3-5 m/s. With higher speeds the turbine starts generating power depending on the known equation (8) of section 2.2.3.

This will be so until it reaches the nominal power. At this point the turbine activates its regulation mechanisms to maintain the same output power. At very high wind speeds the turbine stops in order to avoid any damage. This stop wind speed is called the cut out wind speed.

### 2.4 Wind turbine classification
According to the most of the authors [12], [15] and [17] wind turbines can be classified by three parameters: the direction of the rotor axis, the number of rotor blades and the rotor position.

### 2.4.1 Horizontal or vertical axes classification

Vertical axis wind turbines are the machines where drag force causes the torque in the perpendicular direction of the rotation axis. The basic theoretical advantages of a vertical axis turbines are [15]:

- The possibility to place the generator, gearbox etc. on the ground avoiding a tower for the machine.
- Do not need a yaw mechanism to turn the rotor against the wind.
- Vertical axes machine does not needs regulation with wind speed variations since it is self-regulated at high wind speeds

The basic disadvantages are:

- The machine is not self-starting.
- The overall efficiency of the vertical axis machines is usually worst than horizontal axes machines.
- To replace the main bearing for the rotor, it requires removing the rotor on both horizontal and vertical axis machines. But, in the case of vertical axes machine, this means tearing the whole machine down.

Today, all grid-connected commercial wind turbines are built with a propeller-type rotor on a horizontal axis (i.e. a horizontal main shaft), Figure 2.9.

Figure 2.9 Commercial wind turbine with horizontal axis.

## 2.4.2 Classification by the number of blades

A wind turbine does not give more power with more blades. If the machines are well designed, the harvested power is more or less the same with different number of blades [17].

Wind turbines do not harvest power from the aerodynamic resistance; they do from the blades shape. So, the difference between two wind turbines with a different number of blades is the torque generated by each blade and consequently, the rotational speed of the rotor. Besides, wind turbines with multiple blades starts working at low wind speeds, due to their high start-up torque.

A rotor with an odd number of blades (and at least three blades) can be considered as a disk when calculating the dynamic properties of the machine.

In the other hand, a rotor with an even number of blades will give stability problems for a machine with a stiff structure. At the very moment when the uppermost blade bends backwards, because it gets the maximum power from the wind, the lowermost blade gets the minimum energy from the wind, which generates mechanic stress to the structure. Thus most of the modern wind turbines have three blades [12].

## 2.4.3 Upwind or downwind classification

In this classification the machines can be upwind or downwind depending on the position of the rotor, Figure 2.10. Upwind machines have the rotor facing the wind, on the contrary downwind machines have the rotor placed on the lee side of the tower.

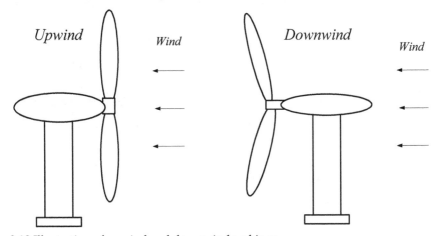

Figure 2.10 Illustration of upwind and downwind turbines.

Downwind machines have the theoretical advantage that they may be built without a yaw mechanism. If the rotor and the nacelle have a suitable design that makes the nacelle follows the wind passively. Another advantage is that the rotor may be made using more flexible materials. Thus, the blades will bend at high wind speeds, taking part of the load off the tower. Therefore, downwind machines may be built somewhat lighter than upwind machines.

The main drawback on downwind machines is that they are influenced by the wind shade behind the tower. When blades cross the wind shade behind the tower, they lose torque and get it back again, causing periodic effort variations in the rotor [17]. Therefore, by far the vast majority of wind turbines have upwind design.

## 2.5 Wind turbines

### 2.5.1 Wind turbine components
A general outline of the components of a wind turbine is given by the following figure:

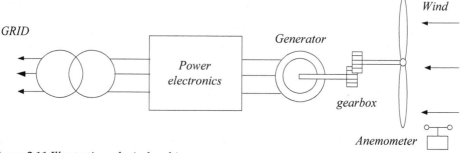

Figure 2.11 Illustration of wind turbine components.

**Rotor blades:** Device to harvest the energy for the wind. At this part the kinetic energy of the wind is transformed into a mechanical torque.

**Anemometer and wind vane**: Devices for measuring wind speed and direction.

**Gearbox:** To convert the slowly rotating, high torque power from the wind turbine rotor to a high speed, low torque power rotation.

**Electrical generator**: Device to transform mechanical energy into electrical energy.

**Associated power electronics:** The part of the wind turbine where electric power is adapted to the frequency and the voltage amplitude of the grid.

**Transformer:** The turbines have their own transformer to step-up the voltage level of the wind turbine to the medium voltage line.

### 2.5.2 Electric generator
An electric generator converts mechanical energy into electrical energy. Synchronous generators are used in most traditional generators (hydro, thermal, nuclear ...). But if these kinds of generators are directly connected to the main grid, they must have fixed rotational speed in synchronism to the frequency of the grid. Thus, torque fluctuations in the rotor (like the fluctuations caused by the wind speed variations) are propagated through the machine to the output electric power.

Furthermore, with fixed speed of the rotor, the turbine cannot vary the rotational speed in order to achieve the optimum speed and extract the maximum torque from the wind. So, with fixed speed the aerodynamic losses are bigger.

Due to these drawbacks, synchronous generators are only used in wind turbines with indirect grid connection. The synchronous generator is controlled electronically (using an inverter), as a result the frequency of the alternating current in the stator of the generator may be varied. In this way, it is possible to run the turbine at variable rotational speed. Consequently, the turbine will generate alternating current at exactly the variable frequency applied to the stator.

On the other hand, asynchronous generators can be used directly or indirectly connected to the grid. Due to the fact that this kind of generators allows speed variations (little) when is connected directly to the grid. Hence, until the present day, most wind turbines in the world connected directly to the grid use a so-called three phase asynchronous generator (also called induction generator) to generate electric power.

### 2.5.3 Wind turbine systems

### 2.5.3.1 Fixed Speed (one or two speeds)

Introduced and widely used in the 80s, the concept is based on a 'squirrel cage' asynchronous generator (SCIG), the rotor is driven by the turbine and its stator is directly connected to the grid. Its rotation speed can only vary slightly (between 1% and 2%), which is almost "fixed speed" in comparison with other wind turbine concepts. So, as its name says, this type of generators cannot vary the speed of the turbine to the optimum speed and extract the maximum torque from the wind.

Aerodynamic control is mostly performed using passive stall, and as a result only a few active control options can be implemented in this kind of wind turbines.

SCIGs directly connected to the grid do not have the capability of independent control of active and reactive power, therefore, the reactive power control is performed usually by mechanically switched capacitors.

Their great advantage is their simple and robust construction, which leads to lower capital cost. In contrast to other generator topologies, FSIGs (Fixed Speed Induction Generators) offer no inherent means of torque oscillation damping which places greater burden and cost on their gearbox.

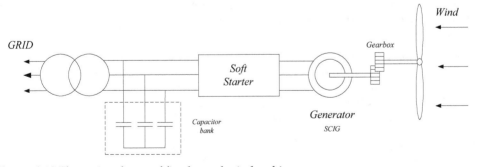

Figure 2.12 The main scheme of fixed speed wind turbine.

The concept exists in both single and double speed versions. The double speed operation gives an improved performance and lower noise production at low wind speeds [18].

European market share: 30% (2005)

Manufacturers: Suzlon, Nordex, Siemens Bonus, Ecotecnia. [18].

### 2.5.3.2 Limited variable speed

Limited variable Speed wind turbines used by Vestas in the 80s and 90s are equipped with a 'wound rotor' induction generator (WRIG). Power electronics are applied to control the rotor electrical resistance, which allows both the rotor and the generator to vary their speed up and down to ± 10% [18].

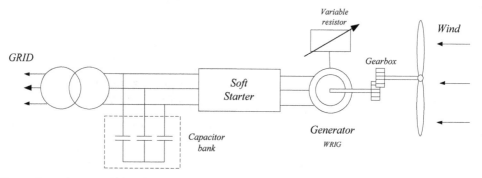

Figure 2.13 The main scheme of limited variable speed wind turbine.

European market share: 10% (2005)

Manufacturers: Vestas (V27, V34, V47). [18].

### 2.5.3.3 Improved variable Speed with DFIG

This system combines advantages of previous systems with advances in power electronics. The DFIG is a wound rotor induction generator whose rotor is connected through frequency converter. In the other hand, stator is directly connected to the grid. As a result of the use of the frequency converter, the grid frequency is decoupled from the mechanical speed of the machine allowing a variable speed operation. Thus maximum absorption of wind power is possible.

Approximately 30% - 40% of the output power goes through the inverter to the grid, the other part goes directly through the stator. The speed variations window is approximately 40% up and down from synchronous speed. The application of power electronics also provides control of active and reactive power, i.e. the DFIG wind turbine has the capability to control independently active and reactive power.

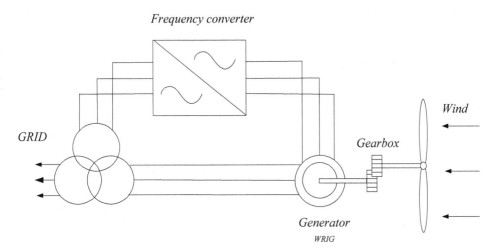

*Frequency converter*

Figure 2.14 The main scheme of improved variable speed with DFIG wind turbine.

European market share: 45% (2005)

Manufacturers: General Electric (series 1.5 y 3.6), Repower, Vestas, Nordex, Gamesa, Alstom, Ecotecnia, Ingetur, Suzlon. [18].

### 2.5.3.4 Variable Speed with full scale frequency converter

The stator of the generator is connected to the grid through a full-power electronic converter. Various types of generators are being used: SCIG, WRIG (Wound rotor induction generator), PMSG (permanent magnet synchronous generator) or WRSG (wound rotor synchronous generator). The rotor has excitation windings or permanent magnets. Being completely decoupled from the grid, it can provide even a more wide range of operating speeds than DFIGs. This kind of wind turbines has two variants: direct drive and with gearbox.

The basic theoretical characteristics of a variable speed with full scale frequency converters are [19]:

> • The DC link decouples completely the generator from the Grid. As the grid frequency is completely decoupled, the generator can work at any rotational speed. Besides changes in grid voltage does not affect the dynamics of the generator.
> • The converters have equal rated power as the generator does, not 30% - 40% like DFIG wind turbines.
> • The converters have full control over the generator.
> • This kind of wind turbine provides complete control over active and reactive power exchanged with the grid. Moreover, it is possible to control the voltage and reactive power in the grid without affecting the dynamics in the generator. As long as there is not a grid fault.

European market share: 15% (2005)

Manufacturers: Enercon, MEG (Multibrid M5000), GE (2.x series), Zephyros, Winwind, Siemens (2.3 MW), Made, Leitner, Mtorres, Jeumont. [18].

### 2.5.3.4.1 Full scale frequency converter with gearbox

The generator uses a two stage gearbox to connect the low-speed shaft to the high-speed shaft, with all the problems associated to the gearbox, like the maintenance or the torque losses.

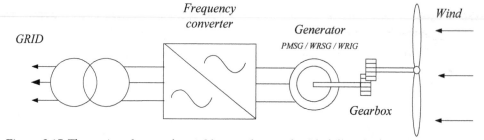

Figure 2.15 The main scheme of variable speed geared with full scale frequency converter wind turbine.

### 2.5.3.4.2 Full scale frequency converter with direct drive

This kind of solutions avoids the gearbox and brushes, so, the implementation of the direct drive in a wind turbine improves the mechanic reliability and produces less noise.

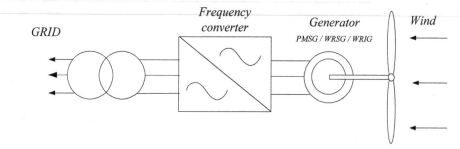

Figure 2.16 The main scheme of variable speed direct drive with full scale frequency converter wind turbine.

### 2.5.4 Active power control
<u>Pitch controlled</u>

On a pitch controlled wind turbine, an electronic controller checks the output power of the turbine several times per second. If the output power is bigger than the rated power, it sends an order to the blade pitch mechanism to pitches (turns) the rotor blades out of the wind. In the other hand, if the output power is lower than the rated power the blades are turned back into the wind, in order to harvest the maximum energy.

Stall controlled

The stall controlled wind turbines are regulated by the aerodynamic loss in the blades. The geometry of the rotor blade profile is aerodynamically designed, to create turbulences on the side of the rotor blade which is not facing the wind, at the moment which the wind speed becomes too high. In this way, it is possible to waste the excess energy in the wind.

Control using ailerons (flaps)

Some older wind turbines use ailerons (flaps) to control the power of the rotor, just like aircraft use flaps to alter the geometry of the wings. But mechanical stress caused by the use of these flaps can damage the structure. Therefore this kind of control only is used in low power generators.

## 2.6 Offshore wind energy vs onshore wind energy
Offshore wind energy in comparison with onshore wind energy has the following advantages / disadvantages [20], [21]:

Advantages:

- **Bigger resource.**
- **Less roughness.**
- **Easier to transport big structures.**
- **Less environmental impact.**

**Bigger resource:** Winds are typically stronger at sea than on land. In the European wind atlas is clearly shown how the wind resource is more abundant in the sea Figure 2.17.

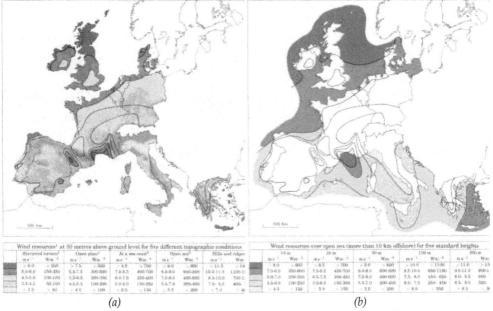

(a)                                                                (b)

Figure 2.17 The average wind speed in Europe, in land (a) and offshore (b) [22].

Besides, the sea has huge spaces to place wind turbines, thus it is possible to install much larger wind farms than in land. The Arklow Bank wind farm has plans to expand its rated power to 520 MW and in Germany and France are proposals to create wind farms with over 1,000 MW.

**Less roughness:** At sea the roughness is lower than in land. As seen in section 2.1.1, the power coefficient (alpha) is much smaller and wind power potential at the same height (equation (2)) is bigger. Moreover, the wind at sea is less turbulent than on land, as a result, wind turbines located at sea may therefore be expected to have a longer lifetime than land based turbines.

In the same way, at sea there are not obstacles to disturb the wind. Consequently, it is possible to build wind turbines with smaller towers, only the sum between the maximum height of the expected wave and the rotor radius.

**Easier to transport big structures:** To transport very large turbine components from the place of manufacturing by road to installation sites on land are logistical difficulties. However, the pieces for offshore wind farms are easily transported by special vessels called Jack-ups.

**Less environmental impact:** Offshore wind farms are too far from the populated areas and they do not have visual impact. Thus they have less noise restrictions than in land, making possible higher speeds for the blade. As a result, it is possible a weight reduction of the blades and mechanical structures, achieving a significant reduction in manufacturing cost.

On the other hand, offshore wind farms present the following disadvantages in comparison with onshore wind farms:

Disadvantages:

- **Operation and maintenance more complicated than in land.**
- **Corrosive environment.**
- **Bigger invest cost.**
- **The energy transmission system to shore.**
- **The depth of the seabed.**

**Operation and maintenance more complicated than in land:** It is not easy access to a facility installed many kilometers into the sea. Therefore it's more complicated the ensemble and maintenance of the facility.

**Corrosive environment:** At sea the salinity and humidity increases the corrosion rate of materials.

**Bigger investment cost:** The cost of the foundations and the transmission system of these facilities is more expensive than onshore wind farms. So the cost per MW installed offshore is about 2.5 times bigger than the cost of installed MW in land.

**The energy transmission system to shore:** The electrical facilities to connect the areas with big offshore wind energy potential with the energy consumption areas are not prepared to transport huge amount of energy.

**The depth of the seabed:** The cost and construction difficulties for an offshore wind farm increases with the water depth.

## 2.7 Chapter conclusions

Offshore wind presents great advantages to develop wind energy, due to the fact that it has a high potential that today still remains largely untapped. However, the opportunities for advancing offshore wind technologies are accompanied by significant challenges, such as: the exposure of the components to more extreme open ocean conditions, the long distance electrical transmission systems on high-voltage submarine cables or turbine maintenance at sea.

Despite of those technological challenges, also have significant advantages. Turbine blades can be much larger without land-based transportation / construction constraints and the blades also are allowed to rotate faster offshore (no noise constraints), so at sea can be installed wind turbines with higher rated powers. Furthermore, the wind at sea is less turbulent than on land.

Thus, the bigger capital costs (twice as high as land-based) can be partially compensated by the higher energy of the wind at sea. In this way, in recent years the average rated power of installed new offshore wind farms has been multiplied by 15. In conclusion, offshore wind is a real opportunity to develop wind energy in the upcoming years.

# Chapter 3

# Offshore Wind Farms

In this chapter an overview of the current technology of the offshore wind farms is performed. This survey is focused into the two main parts of the offshore wind farms electric connection infrastructure: the energy collector system (the inter-turbine medium voltage grid) and the energy transmission system, which are separately evaluated in the present chapter.

Firstly, the AC and DC transmission options to carry the energy from the offshore wind farm to the main grid are described and then, a discussion about the advantages /disadvantages of those AC and DC transmission options is performed. The discussion about the best transmission option is based on the rated power of the wind farms and their distance to shore.

As for the energy transmission system, for the energy collector system of the wind farm, the different configuration options are described. However, for the energy collector grid only AC configurations are taken into account.

In this way, the spatial disposition of the wind turbines inside the inter-turbine grid, the cable length between two wind turbines or the redundant connections of the inter-turbine grid are analyzed.

## 3.1 Historic overview of offshore wind farms
The fast growth of the onshore wind power in Europe, a small and populated area, has led to a situation where the best places to build a wind farm onshore are already in use. However, in the sea, there is not a space constraint and it is possible to continue installing wind power capacity.

The first country to install an offshore wind farm was Denmark in 1991. In the same decade, Netherlands also installed some wind farms very close to shore. So, offshore wind farms are a recently developed technology.

At the early 90s they were very little 6 MW of average rated power, built in very low water depths and with small wind turbines.

However, after this first steps, offshore wind farms are being installed in deeper and deeper waters. Thus, at the end of the 2000s the average water depth of new wind farms multiplied by three, see Figure 3.1.

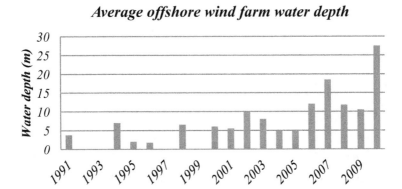

Figure 3.1 Evolution of the average offshore wind farms water depth [23].

But, not only is increasing the average water depth of the wind farms. As the developers are gaining experience / technology in this field and there are more constructed examples. The wind farms are being constructed with bigger rated powers and at locations with longer distances to shore. Thus, from 90s to the next decade the average rated power of installed new offshore wind farms has been multiplied by 15, see Figure 3.3. In parallel with the growth of the average wind farms capacity, the average distance to shore of the wind farms increases as well, see Figure 3.2.

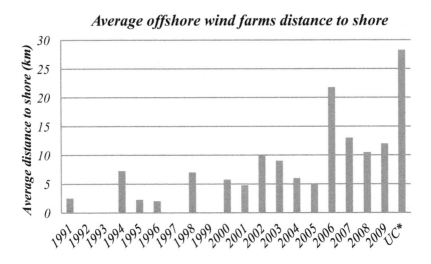

Figure 3.2 Evolution of the average offshore wind farms distance to shore in km [23].

## *Offshore wind farms average capacity*

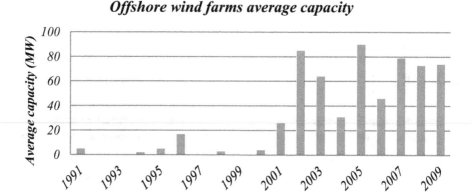

Figure 3.3 Evolution of the offshore wind farms average capacity in MW [23].

As a result, all the biggest offshore wind farms currently in operation are been opened the last few years. Furthermore, four of the five biggest offshore wind farms have been opened during 2010, see Table 3.1

| Name | Country | Year | N° of turbines | Length to shore (Km) | Rated power (MW) |
|---|---|---|---|---|---|
| Thanet | UK | 2010 | 100 | 7.75 | 300 |
| Horns Rev 2 | Denmark | 2009 | 91 | 30 | 209.3 |
| Nysted II/ Rødsand II | Denmark | 2010 | 90 | 23 | 207 |
| Robin Rigg | UK | 2010 | 60 | 9.5 | 180 |
| Gunfleet Sands | UK | 2010 | 48 | 7 | 172.8 |
| Nysted / Rødsand 1 | Denmark | 2003 | 72 | 8 | 165.6 |
| Belwind phase 1 | Belgium | 2010 | 55 | 48,5 | 165 |
| Horns Rev 1 | Denmark | 2002 | 80 | 14 | 160 |
| Prinses Amalia | Netherlands | 2008 | 60 | 23 | 120 |
| Lillgrund | Sweden | 2007 | 48 | 10 | 110.4 |
| Egmondaan Zee | Netherlands | 2007 | 36 | 10 | 108 |
| Inner Dowsing | UK | 2008 | 27 | 5 | 97.2 |
| Lynn | UK | 2008 | 27 | 5.2 | 97.2 |

Table 3.1 Biggest constructed offshore wind farms in EU.

Despite those examples, today the most of the offshore wind farms have a relatively small capacity (<60 MW) and are located relatively close to shore (less than 20 Km), but as is been listed before, also pretty huge wind farms (100-300MW) have been built al locations far away into the sea (> 45 Km).

The new offshore wind farms and offshore wind farm projects are bigger and bigger and at longer distances. As can be seen in Figure 3.4 based on the characteristics of built offshore wind farms which summarize the previous Figure 3.2 and Figure 3.3.

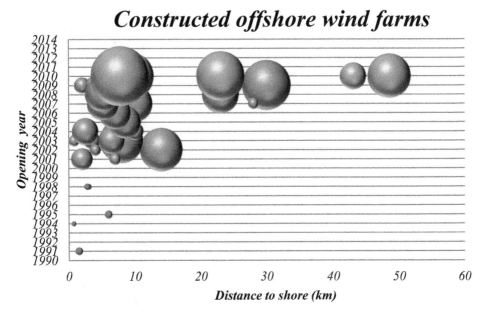

Figure 3.4 Constructed offshore wind farms, rated power (size of the bubbles) depending on the distance to shore and opening year. Offshore wind farms opened in 90s (Red bubbles) and offshore wind farms opened in 2000s (blue bubbles).

Furthermore, according to [24] this trend will continue or will increase in upcoming years, as shown in Table 3.2.

|  | 90s | 2000s | 2010-2030 |
| --- | --- | --- | --- |
| Countries with offshore wind | 3 | 7 | 20+ |
| Average wind farm size | 6 MW | 90 MW | >500 MW |
| Average yearly installed capacity | 3 MW | 230 MW | 6000 MW |
| Average turbine size | < 0.5 MW | 3 MW | 5-6 MW |
| Average rotor diameter | 37 m | 98 m | 125-130 m |
| Average water depth | 5 m | 15 m | >30 m |

Table 3.2 Evolution of the offshore wind energy and future trend.

## 3.2 Offshore wind farms energy transmission system

The key difference between onshore wind farms and offshore wind farms is the different environment of their locations. As a result, offshore wind farms must be provided with submarine cables for the energy transmission.

On the contrary of overhead power cables, subsea power cables have a high capacitive shunt component due to their structure [25]. When a voltage is applied onto a shunt capacitance, capacitive charging currents are generated. These charging currents increase the overall current of the cable reducing the power transfer capability of the cable (which is thermally limited). Therefore, the power transfer capability for a specific cable decreases depending on its shunt capacitance (see section 4.3).

Like the capacitors, the shunt capacitive component of the cables generates more reactive power and charging currents depending on three factors:

- The length (magnitude of the shunt capacitive component).

- The applied voltage.

- The frequency of the applied voltage.

The length is determined by the location of the wind farm and it cannot have big changes. The transmission voltage is directly related to the current and the wind farms rated power. Thus, the main variable which can be changed is the frequency.

Consequently, there are two different types of transmission system configurations: AC and DC. In DC (zero frequency) there are not charging reactive currents, but the energy distribution, the energy consumption and the energy generation is AC voltage.

Therefore, the main drawback of the DC configurations is that they need to transform the energy from AC to DC and vice versa. So, until now, any offshore wind farm with HVDC transmission system has been build. However, a lot of studies are been conducted in this field.

### 3.2.1 AC Configurations

AC cable systems are a well understood, mature technology. For this reason, all the built wind farms up to now have an AC transmission system to connect the wind turbines to the distribution grid. The distribution grid and the generators are AC, thus, DC/AC converters to transform the evacuated energy are not necessary.

With regards to different types of AC configurations, the different options are divided into two families: HVAC (High Voltage AC) transmission systems and MVAC (Medium Voltage AC) transmission systems.

HVAC transmission system has a local medium voltage wind farm grid (20-30kV) connected to a transformer and a high voltage transmission system. Thus the transmission system requires an offshore platform for the step-up transformer. On the contrary, in MVAC configurations the local medium voltage wind farm grid is used both for connecting all wind turbines and to evacuate the generated power.

As discussed before, AC cables have limited their power transfer capability by the length. But this fact does not mean that wind farms power transfer capability must be limited. If one three-phase connection cannot evacuate the rated power to the required length, is possible to use multiple three-phase connections. For example, Kriegers flak offshore wind farms have planned a transmission system with multiple HVAC connections [26].

On the other hand, for medium voltage configurations (according to [27]), the maximum practical conductor size for operation at 33 kV appears to be 300 mm2, giving a cluster rating in the range of 25 to 30 MW. For wind farms with bigger rated powers more three-phase connections are used. Wind farm is divided into clusters and each cluster is fed by its own, 3 core, cable from shore. At the same voltage level of the wind farms local inter-turbine grid 24-36kV [28].

Therefore, an AC configuration presents multiple design options depending on the transmission voltage level and the number of three-phase connections. In this work, 3 different AC-systems are investigated: HVAC, MVAC and multiple HVAC.

### 3.2.1.1 High voltage AC transmission HVAC

The first configuration to be discussed is the HVAC. This lay-out is commonly used by large offshore wind farms such as: Barrow - 90MW, Nysted - 158MW or Horns Rev - 160MW.

To transmit the energy produced in the wind turbines to the point where the electric grid is strong enough to absorb it, the HVAC transmission systems follows roughly the same lines for the grid connection scheme. A typical layout for such a scheme is depicted in Figure 3.5.

As shown, the wind turbines are connected to a medium voltage local inter-turbine grid, where the energy of the wind farm is collected to be transmitted to shore, beyond is the transmission system.

The transmission system is made up by: an offshore substation (step-up transformer), submarine cables, the interface between the wind farm and the point of common coupling (substation) and the main grid.

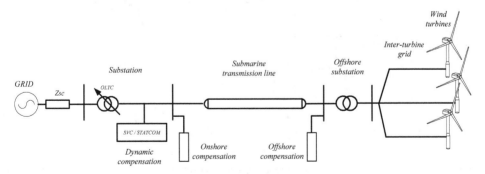

Figure 3.5 Typical layout of HVAC transmission system.

The inter-turbine network is extended from each wind turbine to the collecting point, and due to the fact that it may have a length of kilometers [ 10 ], is typically medium voltage 20-36kV ( Horns Rev 24kV, Barrow 33kV or Nysted 34kV).

In the collecting point the voltage is increased to the required level in the transmission system (Barrow 132kV, Nysted 132kV, Horns Rev 150kV). The energy is then transmitted from the wind farm to the grid interface (substation) over the transmission system. The substation adapts the voltage, frequency and the reactive power of the transmission system to the voltage level, frequency and reactive power required by the main grid (in the PCC) in order to integrate the energy.

This configuration only has one electric three-phase connection to shore, consequently, if this does not work, the whole offshore wind farm is disconnected.

### 3.2.1.2 Multiple MVAC

In this configuration the wind farm is divided into smaller clusters and each cluster is connected by its own three-phase cable to shore. This connection is made in medium voltage, at the same voltage level of the wind farms local inter-turbine grid 24-36kV. Therefore, due to the fact that the voltage level of the transmission system and the inter-turbine grid is the same, this electrical configuration avoids the offshore platform (where is placed the step-up transformer) and its cost.

MVAC electrical configurations are used by small wind farms located near to the shore. For example: Middelgrunden (30kV-40MW) at 3Km to shore, Scroby Sands (33kV-60MW) located 2.3Km seaward o North Hoyle (33kV -60MW) at 7-8Km offshore.

The lay-out of the grid connection scheme for an MVAC transmission system is shown in Figure 3.6

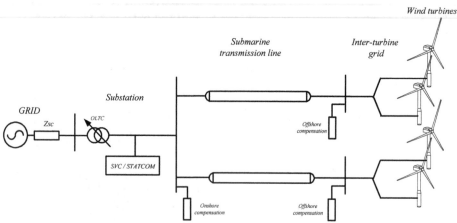

Figure 3.6 Typical layout of multiple MVAC transmission system.

MVAC configuration avoids offshore substantiation and its associated cost, but the energy transmission is through more than one three-phase connection and with lower voltage level than in HVAC configurations. As a result, the energy transmission needs to be made with more current and it is also possible to increase the conduction losses.

As an example, in equations (9)-(14) are compared MVAC configurations conduction losses with HVAC configurations conduction losses for the same cable resistivity and the same transmitted energy (without considering the charging currents due to the capacitive component). The compared cases are 150kV HVAC configuration and a 30kV MVAC configuration with two connections to shore.

$$P = \text{Constant} \; ; \; P_{HVAC} = P \; ; \; 2 \cdot P_{MVAC} = P \; ; \; R_{cable} = \text{Constant} \tag{9}$$

$$V_{HVAC} = 150kV \; ; V_{MVAC} = 30kV \tag{10}$$

$$V_{HVAC} = 5 \cdot V_{MVAC} \tag{11}$$

$$P = V \cdot I \tag{12}$$

$$I_{MVAC} = \left(\frac{5}{2}\right) \cdot I_{MVAC} \tag{13}$$

$$P_{loss} = I_{active}^2 \cdot R \tag{14}$$

$$P_{loss\_HVAC} = I_{HVAC}^2 \cdot R \tag{15}$$

$$P_{loss\_MVAC} = 2 \cdot \left(I_{MVAC}^2 \cdot R\right) = 2 \cdot \left(\left(\frac{5}{2}\right)^2 \cdot I_{HVAC}^2 \cdot R\right) \tag{16}$$

Where: $P$ = Rated power of the offshore wind farm, $P_{HVAC}$ = Rated power of the HVAC three-phase connection, $P_{MVAC}$ = Rated power of each MVAC three-phase connection, $V_{HVAC}$ = HVAC connections voltage level, $V_{MVAC}$ = MVAC connections voltage level, $I_{HVAC}$ = The necessary active current to transmit the rated power of the three-phase HVAC connection at the rated transmission voltage, $I_{MVAC}$ = The necessary active current to transmit the rated power of the three-phase MVAC connection at the rated transmission voltage, $P_{loss}$ = Conduction active power losses, $P_{loss\_HVAC}$ = Conduction active power losses for HVAC connection and $P_{loss\_MVAC}$ = Conduction active power losses for MVAC connection.

The calculated conduction losses are referring only to the active current, thus to calculate total conduction losses, the conduction losses produced by reactive power (charge/discharge currents) must be taken into account.

MVAC configurations have less voltage (MV) than HVAC configurations (HV), as a result, the reactive power/charging currents generated in submarine cables are lower.

Depending on the transmitted reactive power, the conduction loses increases. Therefore, in cases with big capacitive shunt component of the cable (which depends on the length and cable characteristic) high transmission voltage levels increases significantly reactive currents. Reached such a point where high voltage three-phase connections have more conduction losses than medium voltage three-phase connections

$$|I| = \sqrt{I_{active}^2 + I_{reactive}^2} \qquad (17)$$

In this type of configurations with more than one three-phase connections to shore, if a fault occurs in one of those connections, the faulted clusters can remain connected sharing another connection with other cluster, while the faulty cable is repaired. In this way the reliability of the wind farm is improved.

These redundant connection/s are important when around the location of the wind farm is a lot of marine traffic which can damage submarine cables or if the wind farm location has an extreme climate in winter which makes impossible to repair a cable on this year station [29].

### 3.2.1.3 Multiple HVAC

This electrical configuration is a combination of the previous two configurations. The wind farm is divided into smaller clusters but these clusters are not connected directly to shore. In this case, each cluster has a collector point and step-up transformer. At this point the voltage is increased to the required level in the transmission system. Typical layout for these configurations is depicted in Figure 3.7.

Offshore wind farms with multiple HVAC transmission systems have large rated power, for example Kriegers Flak (640MW) or Robin Rigg (East/west – 180MW). These wind farms are under construction.

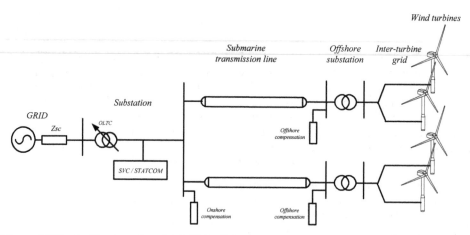

Figure 3.7 Typical layout of multiple HVAC transmission system.

In this electrical configuration, the wind farm is divided into clusters and the rated power of these clusters is a design option. In the same way, the transmitted power through a three-phase connection which depends on the number of clusters can be selected by design also. Moreover, the transmission voltage can be different to the local inter-turbine grids voltage, making possible multiple and different combinations.

In this type of configurations also are more than one three-phase connections to shore. As a result, it is possible to make redundant connections and if a fault occurs in one of those connections, the wind farm does not have to disconnect.

With regards to the drawbacks, this electrical configuration needs several offshore platforms (substations) to increase the voltage to the required level by the transmission system. Another option is place several transformers in the same platform, but the size of the platform increases. Thus, in both cases the cost of the platform increases.

### 3.2.2 DC Configurations

DC configurations do not generate charging currents or reactive power due to the capacitive shunt component of the submarine cable. This is a huge advantage in comparison with AC configurations, but they also have a big drawback: The distribution grid, the energy consumption and the generators are AC voltages. Therefore these configurations must be provided with AC/DC converters to adequate the voltage to the energy transmission.

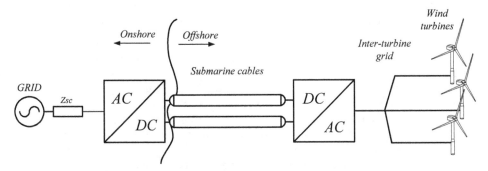

Figure 3.8 Generic layout of bipolar HVDC transmission system.

The AC/DC converters technology and the transmission voltage characteristics are the main features of this kind of configurations:

The transmission can be through mono-polar voltage (Using a single cable) with return to earth or bipolar (Using two cables), Figure 3.8. Due to the extra cable the transmission system can evacuate more power to shore and improve the reliability of the system providing redundancy.

With regards to the AC/DC converters technology, these can be LCC (Line commutated converters) based on thyristors or VSC (Voltage source converter) based on switching devices with the capability to control their turn on and off.

Line Commutated Converter (LCC) devices have been installed in many power transmission systems around the world, so it is a mature technology.

In a line commutated converter, it is possible to control the turn on instant of the thyristor, but the turn off cannot be controlled. Therefore, systems based on this technology for the converters are more susceptible to potential AC grid faults than VSC converters.

In addition, a LCC converter needs a minimum reactive power to work (a minimum current), consequently, this type of converters need voltage on both sides (offshore and onshore) to start working.

Voltage Source Converter (VSC) solution is comparatively new compared to the LCC solution. As the main advantage, the semiconductors switching is decoupled to the grid voltage. Thus, the VSC solutions are able to supply and absorb reactive power to the system independently and may help to support power system stability. As a result, VSCs are suitable for systems with low short circuit power.

In VSC configurations the substation requires fewer components to filter, due to this task can be performed by the converter itself. On the negative side of these converters is their high cost (higher than LCC converters).

### 3.2.3 AC vs. DC Configurations
To date, any of the built offshore wind farms have a HVDC transmission system, but a lot of studies are been conducted in this field. Furthermore, a small-scale demonstration of HVDC technology is working successfully in Tjaereborg Enge (Denmark). This wind farm has four wind turbines of different types with total rated capacity of 6.5MW.

An AC configuration presents a sort of advantages and disadvantages in comparison with DC configurations. For example, with regard for the energy transmission and integration into the main grid has the following advantages and disadvantages [30]:

- AC configurations do not need to convert all the transmitted energy, due to the generators and the main grids have AC voltage.
- The components for AC configurations are more standard, consequently the cost of the offshore platform is lower.
- AC configurations have well proven and reliable technology. All the built offshore wind farms until now are AC.

In the other hand, the disadvantages for an AC transmission configuration are:

- o Long AC cables generate large amounts of reactive power, due to their high capacitive shunt component.
- o The associated charging currents to the reactive power reduce the transfer capacity of the cable. This reduction is in proportion to the capacitive shunt component, mainly for the cable length
- o All faults in the main power grid affect the collecting AC offshore grid directly and vice versa. So depending on the grid requirements, system may require fast voltage and reactive power control during fault operation.

DC configurations have several features that make them attractive. For example, in these kind of configurations, there are not charging currents. Consequently, the power transfer capability for long cables is not reduced and the reactive power compensation is not needed. As a result, they are more suitable for long distances.

Besides, a DC configuration needs AC/DC converters at both ends of the line. Thus, the transmission system has the capability to control both the voltage and the power injected to the main grid.

These converters provide also electrical decoupling between the collecting AC offshore grid and the main power grid for faults in the main power grid.

On the other hand, there are several disadvantages associated to this technology, mainly the cost:

- These configurations have higher substation cost, both offshore and onshore.
- Higher overall losses (switching losses in power converters)
- Limited offshore experience with VSC-transmission systems
- It is possible to increase the injected harmonic level to the main grid, especially with LCC technology [31], [32].

The advantages and disadvantages of each transmission configuration can be seen summarized in Table 3.3

| AC | DC | |
|---|---|---|
| | LCC | VSC |
| Is possible to avoid offshore platform. | Needs an offshore platform. | Needs an offshore platform. |
| Is possible to avoid switching losses (to avoid converters). | Switching losses at AC/DC converters. | Switching losses at AC/DC converters. |
| Need reactive power compensation | Not need reactive power compensation. | Not need reactive power compensation. |
| All faults in the main power grid affect the collecting AC offshore grid directly and vice versa | Electrical decoupling between the collecting AC offshore grid and the main power grid. | Electrical decoupling between the collecting AC offshore grid and the main power grid. |
| charging currents reduces the power transfer capacity | There are not charging currents | There are not charging currents |
| Power transmission capability in both directions. | To change the direction of the power flow need to change the polarity. | Power transmission capability in both directions. |
| Mature and reliable technology in offshore applications. | Well proven technology but in other applications. | Well proven technology but in other applications. |
| Less cost due to the standard components | Needs a minimum reactive power to work. Also in no-wind conditions | High cost. |

Table 3.3 Comparison of AC and DC transmission systems.

### 3.2.4 The optimum layout depending on rated power and distance

Several studies about which transmission configuration is the optimum depending on the distance to shore and rated power are been analyzed in this chapter [33], [34] and [35].

In [36] the analysis is focused on HVAC and HVDC configurations with pretty high rated powers (until 900MW) and long distances to shore (up to 300 Km). As a conclusion, this survey determine AC technology like the optimum for offshore wind farms with rated powers until 200MW and distances to shore shorter than 100 Km.

However, the survey does not determine the optimum technology in the range between 200MW-400MW with distances to shore between 100Km-250Km. As says in the report AC technology is losing attractiveness increasing the distance to shore. So DC configurations are the optimum option for big power and long distances.

A similar study is presented in [27], which determines the limits of each technology for the transmission system of offshore wind farms as follows: Until 100MW and 100Km to shore the optimum configuration is MVAC, between 100MW-300MW and 100Km-250Km the optimum is HVAC and for bigger rated powers and longer distances HVDC.

Finally in [37], the report places the limits of each technology as follows: The distance limit for using AC transmission configuration is between 80 and 120 km, if the reactive power generated in submarine cables is compensate at both ends. But if the transmission system has a reactive power compensator in the middle of the submarine cable besides the reactive power compensators at the both ends, AC technology may be the optimum until to 180 km. This report considered these limits flexible.

The summarize of the previous studies is shown in Figure 3.9, in this picture are depicted the areas where each type of transmission is the optimum configuration (white) and the areas where depending on the characteristics of each wind farm can be any of them the optimum configuration (striped area).

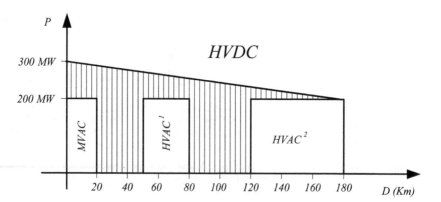

Figure 3.9 Optimum configuration depending on distance to shore and rated power: (1) With reactive compensation at both ends, (2) With reactive compensation at both ends and another one in the middle of the submarine cable.

### 3.3 Offshore wind farms electrical collector system

The local inter-turbine grid can be AC or DC, although this characteristic does not determinate the transmission systems technology. The transmission system can be made with the same technology or not. To date all the inter-turbine grids are AC and a lot of studies of HVDC transmission systems have AC inter-turbine grids. Thus, in the present book, only AC collector grids are considered.

The design of the wind farms collector system begins with the selection of the inter-turbine cable and inter-turbine grids (collection grid) voltage level. The use of voltage levels above 36kV for the inter-turbine grid becomes uneconomic. Due to the impossibility to accommodate switchgear and transformers in each turbine tower, so 33-36kV is widely used for collection schemes [38].

The number of turbines connected to the same cable and the rated power of these wind turbines with the voltage level determines the inter-turbines cable section. This aspect is not trivial, as cables have to be landed on seabed, and a larger section means a larger bending radius (higher stiffness), with consequent difficult maneuvering of cable posing ships and larger mechanical protection components [39].

With regards to the spatial disposition of the wind turbines inside the inter-turbine grid, most offshore wind farms to date, have had simple geometric boundaries and have adopted a straightforward rectangular or rhomboid grid [38].

The cable length between two wind turbines is determined mainly for the aerodynamic efficiency of each turbine. Due to the length between two wind turbines must be enough to avoid turbulences generated at the turbines around it. According to [40] the length between two wind turbines is in the range of 500-1000 m.

In onshore wind farms, the electrical system configuration is usually decided by the turbine and substation positions, and the site track routes. Offshore, on the contrary, to design the inter-turbine grid is more freedom, and at first sight is not clear how to choose from the wide range of possible options [41].

Nevertheless, for wind farm collector systems employed in existing offshore wind farms are various standard arrangements. In this way, four basic designs are identified [40], [34]:
- Radial design.
- Single side radial design.
- Double-sided radial design.
- Star design.


### 3.3.1 Radial design

The most straightforward arrangement of the collector system in a wind farm is a radial design. The wind turbines are connected to a single cable feeder within a string. This design is simple to control and also inexpensive because the total cable length is the smallest to connect all the wind turbines with the collecting point [41]

The radial design is not provided by redundant connections. Consequently if a fault occurs in a cable or at the hub, the entire radial string collapses and all of the wind turbines in the string are disconnected.

The maximum number of wind turbines on each string is determined by the rated power of the generators and the rated power of the submarine cable. The lay-out of this kind of collector systems is shown in Figure 3.10.

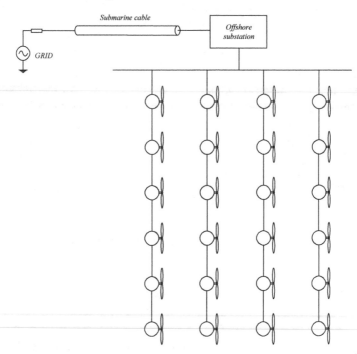

Figure 3.10 Layout of the radial design for local inter-turbine grid.

### 3.3.2 Single-sided ring design

A single-sided ring design is similar to radial design, but with an extra connection. The additional cable connects the last wind turbine with the collector.

In comparison with radial inter-turbine grid design, the additional cable incorporates a redundant path to improve the reliability of the system. Therefore, this additional cable must be able to evacuate all the energy generated in the string.

The main drawback of this lay-out is the required cable length, longer than in the radial design with its associated cost increase.

To justify the use of redundancy in the collecting system, it is considered the energy that will be saved with this redundant cable during its useful life (usually 20 years) and this benefit is confronted with the cost of redundant cable. In this way, In [41] is reported that the most internal power networks of existing offshore wind farms have very little redundancy or none at all.

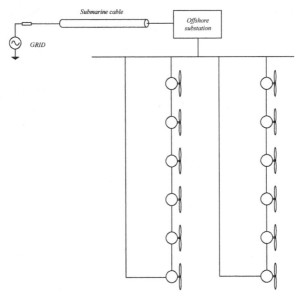

Figure 3.11 Layout of the single sided ring design for local inter-turbine grid.

### 3.3.3 Double-sided ring design

A double-sided design is similar to single-sided ring design but in this case the extra connection is between the last wind turbines of two strings, as is illustrated in Figure 3.12

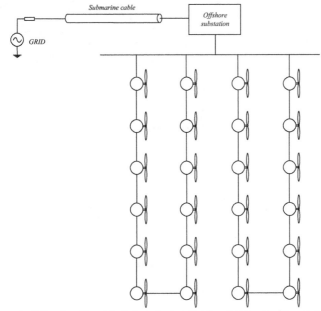

Figure 3.12 Layout of the double sided ring design for local inter-turbine grid.

Connecting the last wind turbines of each string saves cable length, but in the other side, if a fault occurs, the whole output power of two strings is deviated through the same cable. Thus, the inter-turbine cable has to be sized for that purpose.

### 3.3.4 Star design

The star design has a large number of connections, because each turbine is connected directly to the collector point in the center. So this design provides a high level security for the wind farm as a whole. If one cable have a fault, only affects to one turbine (the turbine connected trough this cable).

The additional expense in longer subsea cables is compensated at least in part with less cable sections required by this design. Due to the fact that trough the inter-turbine cable is only transmitted the energy generated by one wind turbine. So the biggest cost implication of this arrangement is the more complex switchgear requirement at the wind turbine in the centre of the star.

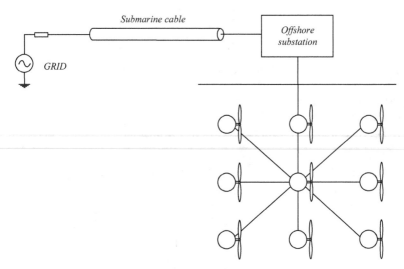

Figure 3.13 Layout of the star design for local inter-turbine grid.

## 3.4 Chapter conclusions

Due to the fact that the energy distribution grid, the energy consumption and the energy generation are AC voltages, all the built offshore wind farms to date have AC transmission system. So, AC alternative is well proven and feasible.

AC and DC configurations have their advantages and disadvantages. All the considered studies in this chapter are agreed that for long-distances to shore DC option would be the optimum if not the only viable. But at present DC transmission options are proposals to adapt the technology to the offshore environment.

The analyzed studies are not agreed about the limits on distance and the rated power for the optimal transmission option (AC or DC). These studies neither are agreed about within each family which transmission configuration (HVAC, MVAC or Multiple HVAC) is the optimum.

In general, the studies are agreed in rough lines. Highlighting that for short distances to shore and low rated power the optimum option is AC and for long distances to shore and big rated power DC. Consequently, depending on the rated power of the wind farm and its features any transmission option (AC or DC) can be the optimum.

Therefore, it is necessary a more detailed analysis, which takes into account features of specific cases in order to define the optimum transmission option and configuration for each offshore wind farm.

# Chapter 4

# Power AC Transmission Lines

The submarine power AC cables have an important role in offshore wind energy. Furthermore, the submarine cables are the main difference between the offshore wind farms transmission system and onshore wind farms transmission system.

Therefore, a proper submarine cable model is crucial to perform accurate evaluations of the offshore wind farms collector and transmission systems. So, in the present chapter the different options to model a submarine cable are evaluated and their accuracy is discussed.

Then based on an accurate and validated submarine cable model, an analysis about the reactive power management in submarine power transmission lines is carried out. Thus, taken into account active power losses, the reactive power generated in the transmission system and the voltage drop for three different reactive power management options, a reactive power compensation option is proposed.

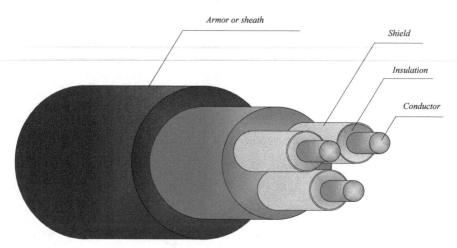

Figure 4.1 Generic representation of a electric power cable.

### 4.1 Basic components of electric power cables

The purpose of a power cable is to carry electricity safely from the power source to different loads. In order to accomplish this goal, the cable is made up with some components or parts. Figure 4.1 shows a description of the cable components, which are:

Conductor: The conductor is referred to the part or parts of the cable which carry the electric power. Electric cables can be made up by one conductor (mono-phase cables), three (three-phase cables), four, etc.

Insulation: Dielectric material layer with the purpose of insulate conductors of different phases or between phases and ground.

Shield: metal coating, which covers the entire length of the cable. It is used to confine the electric field inside the cable and distribute uniformly this field.

Armor or sheath: Layer of heavy duty material used to protect the components of the cable for the external environment.

### 4.1.1 Conductor

Some materials, especially metals, have huge numbers of electrons that can move through the material freely. These materials have the capability to carry electricity from one object to another and are called conductors. Thus, conductor is called to the part or parts of the cable which carry electric power.

The conductor may be solid or made up with various strands twisted together. The strand can be concentric, compressed, compacted, segmental, or annular to achieve desired properties of flexibility, diameter, and current density.

The choice of the material as a conductor depends on: its electrical characteristics (capability to carry electricity), mechanical characteristics (resistance to wear, malleability), the specific use of the conductor and its cost.

The classification of electric conductors depends on the way the conductor is made up. As a result, the conductors can be classified as [42]:

#### 4.1.1.1 Classification by construction characteristics

Solid conductor: Conductor made up with only one conductor strand.

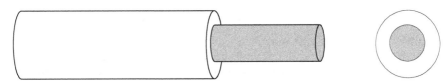

Figure 4.2 Conductor made up with Only one conductor strand.

Strand conductor: Conductor made up with several low section strands twisted together. This kind of conductor has bigger flexibility than solid conductor.

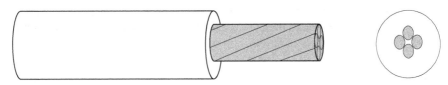

Figure 4.3 Conductor made up with several low section strands twisted together.

### 4.1.1.2 Classification by the number of conductors

<u>Mono-conductor:</u> Conductor with only one conductive element, with insulation and with or without sheath.

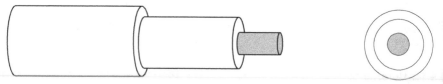

Figure 4.4 Conductor with Only one conductive element.

<u>Multiple-conductor:</u> Conductor with two or more conductive elements, with insulation and with one or more sheaths.

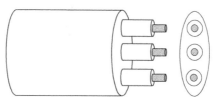

Figure 4.5 Conductor with multiple conductive elements.

### 4.1.2 Insulation

The purpose of the insulation is to prevent the electricity flow through it. So the insulation is used to avoid the conductor get in touch with people, other conductors with different voltages, objects, artifacts or other items.

### 4.1.2.1 Air insulated conductors

A metallic conductor suspended from insulating supports, surrounded by air, and carrying electric power may be considered as the simplest case of an insulated conductor [42].

Air is not a very good insulating material since it has lower voltage breakdown strength than many other insulating materials, but it is low in cost if space is not a constraint. On the contrary, if the space is a constraint, the air is replaced as insulation material for another material with higher voltage breakdown strength [42].

The same occurs in environments where isolation by air is not possible like submarine cables. In this case neither is possible isolation by sea water, since it is not an insulating material.

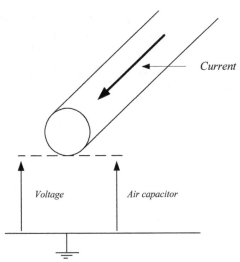

Figure 4.6 Air insulated conductor.

### 4.1.2.2 Insulation by covering the conductor with a dielectric material

In this type of insulation, the conductor is covered by an insulating material with high voltage breakdown strength (a dielectric), usually a polymer.

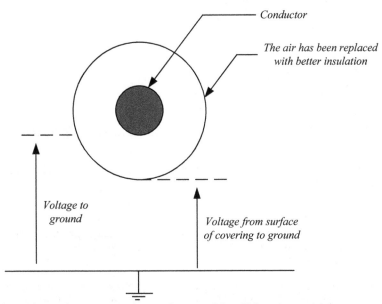

Figure 4.7 Insulation by covering the conductor with a dielectric material.

If the metallic conductor is covered with an insulating material, transmission lines can be placed close to ground or touching the ground. But in this cases when the ground plane is brought close or touches the covering, the electric field lines become increasingly distorted.

Considering the equipotential lines of the electric field, these are bended due to the potential difference on the covering surface. As shown in Figure 4.8.

At low voltages, the effect is negligible. As the voltage increases, the point is reached where the potential gradients are enough to cause current to flow across the surface of the covering. This is commonly known as "tracking." Even though the currents are small, the high surface resistance causes heating to take place which ultimately damages the covering. If this condition is allowed to continue, eventually the erosion may progress to failure [42].

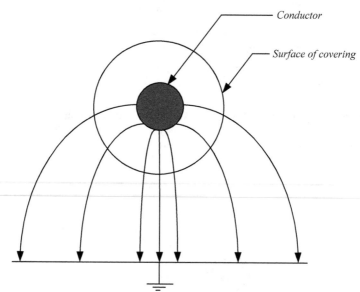

Figure 4.8 Equipotential lines of the conductor's electric field when the transmission line is close to the ground.

Therefore, high voltage power cables close to ground, like submarine cables, are provided with a shield to avoid this effect.

### 4.1.3 The insulation shield
The shield is a metallic coating over the insulation and connected to ground. The purpose of the shield is to create an equipotential surface concentric with the conductor to avoid the bending of the electric field lines.

The shield is also used to avoid the effects of external electric fields on the cable and as a protection for worker staff, through the effective connection to ground. The main reasons to use a shield are:

- To confine the electric field inside the cable between the conductor and the shield.
- To make equal the efforts inside the insulation, minimizing partial electric discharges

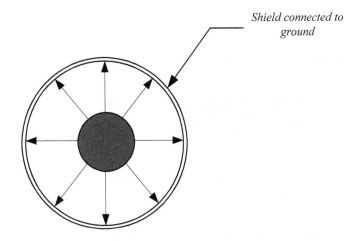

*Shield connected to ground*

Figure 4.9 Equipotential lines of the conductor's electric field when the transmission line is provided with a shield.

- To protect the cable from induction voltages.
- To avoid electromagnetic or electrostatic interferences.

### 4.1.4 Armor or sheath

The purpose of this part of the cable is to protect the integrity of the insulation and the conductor from any mechanical damage such as scrapes, bumps, etc.

If mechanical protections are made by steel, brass or other resistant material, this mechanical protection is called as "armor." The "armor" can be composed by strips, strands or plaited strands.

The armor has especial importance in submarine cables, due to this type of cables are under water and the armor provides mechanical protection against to submarine water currents. Therefore, often submarine cables have an armor made up with a crown of steel strands in order to achieve a good mechanical protection [43].

### 4.2 Power transmission line modeling

### 4.2.1 Power transmission lines electric representation

A power transmission line presents several phenomena. First of all, the conductor of the cable used in power transmission lines has a small resistivity. Resistivity is the scalar property of an electric circuit which determines, for a given current, the rate of which electric energy is converted into heat or radiant energy. The resistivity of a specific cable is given by equation (18).

$$R = \frac{l}{\sigma \cdot A_c} \qquad \text{(ohm)} \qquad (18)$$

Where: '$l$' is the length of the cable, $\sigma$ is the conductivity of the cable and $A_c$ is the conductor area of the cable.

When an electric current flows through a conductor generates a magnetic field around it, which in turn induces an electric field. This field generates a current in the conductor in opposite direction of the original current. This effect is called self-inductance and can be described as [44]:

$$L = \frac{\mu_0}{8\pi} + \frac{\mu_0}{2\pi} \ln\left(\frac{b}{a}\right) \qquad \left(\frac{H}{m}\right) \qquad (19)$$

$$L = 0.5 + 0.2\ln\left(\frac{b}{a}\right) \qquad \left(\frac{mH}{Km}\right) \qquad (20)$$

Where: $\mu_o$ is the magnetic constant or the permeability of the free space and (a, b) are the radius of conductor cylinders (see Figure 4.10).

In case of cables with more than one wire or conducting element, besides the self inductance of each wire, must be also considered the electric field created in other wires. Consequently, the inductance of a multi-conductor cable mainly depends on the thickness of the insulation over the conductors.

Power transmission lines with triangular spatial disposition of the conductors, i.e. with the same separation between the three conductors present a self-inductance given by (21) [45]:

$$L = 0.05 + 0.2\ln\left(\frac{D}{a}\right) \qquad \left(\frac{mH}{Km}\right) \qquad (21)$$

Where: $D$ is the distance between conductors and $a$ is the conductor radius

It is important to highlight that the equations are for power lines with triangular spatial disposition. If the spatial disposition is with conductors in line, the value of the self-inductance is altered.

Another effect to be considered to represent a cable is the capacity of the line to ground (which is represented by the capacitor C). The voltage difference from the conductor to ground causes this effect.

In the cases of cables with insulation and placed close to the ground (like underground or subsea cables), they have to be provided with a shield. Thus, this capacity depends on the dielectric (insulation). Due to the fact that this capacitor represents the capacitive behavior performed between the conductor and the shield (a conductor connected to ground). In the most generic case is calculated by the equation (22).

$$C = \frac{Qe1}{V1-V2} = \frac{Qe2}{V2-V1} \qquad \text{(F)} \qquad (22)$$

Where: $C$ is the capacity of the cable, $V1$ is the voltage of the conductor 1, $V2$ is the voltage of the conductor 2, $Qe1$ is the electric charge stored in the conductor 1 and $Qe2$ is the electric charge stored in the conductor 2.

Simplifying the cable as a cylindrical conductor of radius "$a$" and a cylindrical surface coaxial with the first of radius "$b$" ($a <b$), where the space between them is filled with a dielectric material, Figure 4.10

It is possible to make the assumption that "$a$" and "$b$" (cable cross section) are very small in comparison with length ($l$) of the conductor cylinders (cable). As a result, the length of conductor cylinders (cable) can be considered as infinite, i.e. an ideal cylindrical capacitor. Where its capacity is given by equations (23) and (24) [46]:

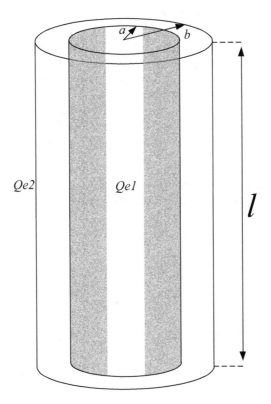

Figure 4.10 Geometrical approximation of the physical form of the cable.

$$C = \frac{2 \cdot \pi \cdot \varepsilon_0 \cdot \varepsilon_r \cdot l}{\ln\left(\frac{b}{a}\right)} \qquad \text{(F)} \qquad\qquad (23)$$

$$C = \frac{\varepsilon_r}{17.97 \cdot \ln\left(\frac{b}{a}\right)} \qquad \text{(μF/Km)} \qquad\qquad (24)$$

Where: $\varepsilon_r$ is the dielectric constant or relative permittivity of the insulating material between conductors, $\varepsilon_0$ is the dielectric constant in the vacuum, $l$ is the length of the conductor cylinders and $(a, b)$ are the radius of conductor cylinders.

Finally, the cable has a leakage current from the conductor to ground (represented by a conductance G). The dielectric is a material with low conductivity, but not zero, i.e. the insulation presents high impedance, nevertheless, this does not mean infinite. Thus, the conductance G represents the current generated from the conductor to ground (the shield connected to ground) through the dielectric because the insulation is not ideal.

In short, transmission lines are basically circuits with distributed parameters, i.e. R, L, C and G are distributed along the whole length of line. Where:

- The distributed resistance R of the conductors is represented by a series resistor (expressed in ohms per unit length).
- The distributed inductance L (due to the magnetic field around the wires, self-inductance, etc.) is represented by a series inductor (henries per unit length).
- The capacitance C between the conductor and the shield is represented by a shunt capacitor C (farads per unit length).
- The conductance G of the dielectric material separating two conductors (the shield and the conductor) is represented by a conductance G, shunted between the signal wire and the return wire (Siemens per unit length).

Therefore, a transmission line can be represented electrically per phase for each differential length as in Figure 4.11 [47], [48] y [49].

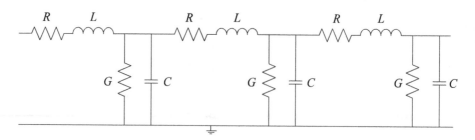

Figure 4.11 Electric representation of the cable per differential length.

In the same way, for three-phase transmission lines, the cable can be represented using three identical schemes per phase as follows, Figure 4.12.

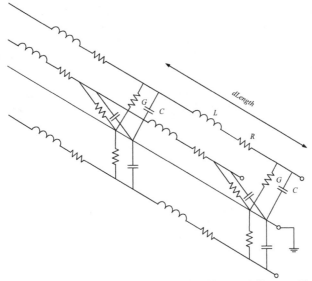

Figure 4.12 Electric representation of the three phase cable per differential length.

### 4.2.1.1 Skin effect

In DC circuits, the current density is similar in all the cross section of the conductor, but in AC circuits, the current density is greater near the outer surface of the conductor. This effect is called skin effect.

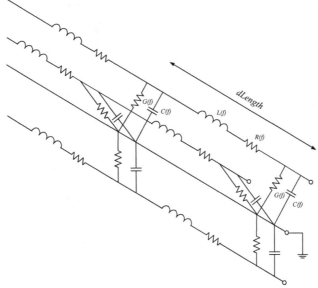

Figure 4.13 Electrical representation of a three-phase cable per differential length with frequency dependent parameters.

Due to this phenomenon, AC resistance of the conductor is greater than DC resistance. Near to the center of the conductor there are more lines of magnetic force than near the rim. This causes an increment in the inductance toward the center and the current tends to crowd toward the outer surface. So at high frequencies the effective cross section area of the conductor decreases and AC resistance increases.

In short, the skin effect causes a variation in the parameters of the cable, due to the non uniform distribution of the current through the cross section of the cable. This variation depends on frequency. Consequently, RGLC parameters are frequency dependent.

If this effect is taken into account the electric representation of the cable for each differential length is represented as shown in Figure 4.13.

### 4.2.2 Power transmission line modeling options

Based on the electric representation of the cables and depending on the cable model requirements, it is possible to perform more or less simplifications, in order to maintain the accuracy of the model and reduce its complexity. Thus, there are several ways for modeling a cable; these models can be classified as follows [50].

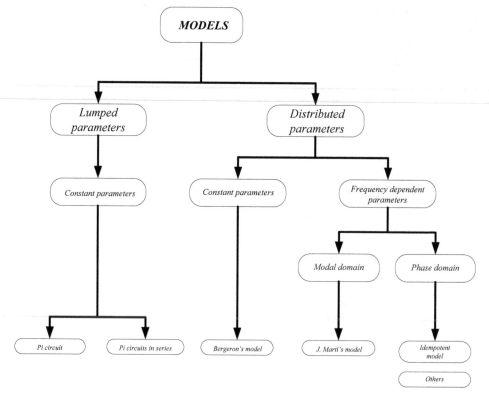

Figure 4.14 Classification of the different types of cable models.

In the present work, these models are divided into models based on constant parameters and models based on frequency dependent parameters. This division is made, because these two groups are based on different electrical representations, Figure 4.12 and Figure 4.13.

Thus, according to [51] for modeling the transmission lines with constant parameters there are the following options:

- Bergeron's traveling wave.
- Standard short, medium and long line models for phasor domain. If only care about 50-60Hz.
- Sequence of single phase "π" segments. In order to model the transients in easy way.

If the objective is the analysis of a wide frequency spectrum accurately, a more accurate model of a line can be developed considering the RGLC parameters distributed and frequency dependent. So, in this report, the following frequency dependent cable models are considered:

- Frequency dependent model in modal domain (J. Marti).
- Frequency dependent model in phase domain (Idempotent model)

### 4.2.2.1 Constant parameter models

All the mathematical analysis of the evolution of the current and voltage to develop constant parameter models are based on a generic segment of the cable with length ($\Delta x$). Therefore, these models are developed starting from the electrical representation of Figure 4.11

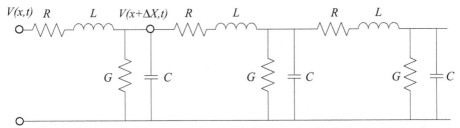

Figure 4.15 Electrical representation of a generic cable segment with constant parameters.

Where: $V(x,t)$ and $V(x+\Delta x,t)$ are instantaneous voltages on x and $x+\Delta x$ respectively and $I(x,t)$ and $I(x+\Delta x,t)$ instantaneous current on $x$ and $x+\Delta x$.

Applying Kirchoff's laws to this circuit, it is possible to obtain the following equations to describe the behavior of the circuit:

$$V(x,t) - V(x + \Delta x,t) = \Delta V \tag{25}$$

$$\Delta V = -(RI(x,t) + L\frac{\partial I}{\partial t})\Delta x \tag{26}$$

$$I(x,t) - I(x + \Delta x,t) = \Delta I \tag{27}$$

$$\Delta I = -(GV(x,t) + C\frac{\partial V}{\partial t})\Delta x \tag{28}$$

In the limit $\Delta x \to 0$, these equations ((26) and (28)) can be expressed as follows:

$$-\frac{\partial V}{\partial x} = RI + L\frac{\partial I}{\partial t} \tag{29}$$

$$-\frac{\partial I}{\partial x} = GV + C\frac{\partial V}{\partial t} \tag{30}$$

The equations (29) and (30) are called the general equations of the transmission lines and all the constant parameter models are based on these equations.

### 4.2.2.2 Standard short, medium and long line models for phasor domain

Taking the derivative of equations (29) - (30), it is possible to obtain the following second-degree ordinary differential equations of the transmission line for voltage and current.

$$\frac{\partial^2 V}{\partial x^2} = \gamma^2 \frac{\partial V}{\partial t} \tag{31}$$

$$\frac{\partial^2 I}{\partial x^2} = \gamma^2 \frac{\partial I}{\partial t} \tag{32}$$

$$\gamma = \alpha_\gamma + i\beta_\gamma = \sqrt{(R + i\omega L)(G + i\omega C)} \tag{33}$$

Where: $\gamma$ is the wave propagation constant, $\alpha_\gamma$ is the real part of the propagation constant which represents the attenuation (Np/m) and $\beta_\gamma$ is the imaginary part of the propagation constant which represents phase velocity (rad/m).

Applying D'Alenbert to the second-degree differential equations, it is possible to obtain the general solution of the equations. This solution consists of two traveling waves that propagate through the line, one from left to right and the other one in reverse.

A simple way to work with traveling voltage waves is representing the system as a two ports network or a quadripole. Moreover, the standard model (for short, medium and large lines) is focused to represent only the steady state of the transmission lines (50-60Hz). Consequently, is possible to use a lumped parameters quadripole.

If the length of the transmission line is small in comparison with the traveling wave length, equation (34). The traveling time of the electromagnetic waves can be neglected, allowing the representation of the transmission system by lumped parameters.

$$\lambda = \frac{v_e}{f_e} = \frac{1}{\sqrt{LC}f_e} \tag{34}$$

Where: $v_e$ is the propagation speed of the traveling wave and $f_e$ is the frequency of the analyzed transient phenomenon

In equation (34) the frequency is inversely proportional to the traveling wave length. As a result, for low frequencies the wave length is large in comparison with the length of the transmission line, i.e. the traveling wave that goes from one node to the other node appears instantly in the second node with virtually no time delay.

Therefore, to obtain the equations of the transmission line for the standard model (50-60Hz) the transmission parameters (also called ABCD-parameters) are used.

On the contrary, if the required frequency for the analysis is high, the traveling wave length is less than the length of the transmission line. As a result, it is not possible to neglect the delay in the wave between the two ends of the cable. In these cases, models based on traveling waves are more accurate [50].

### ABCD parameters

These parameters are based on a lumped parameter two port network or quadripole, Figure 4.16 Each parameter of the ABDC parameters of the two-port network represents:
- (A) the voltage relation between the two ports in open circuit.
- (B) the negative transference impedance in short circuit.
- (C) the transference admittance in open circuit.
- (D) the negative current relation between the two ports in short circuit.

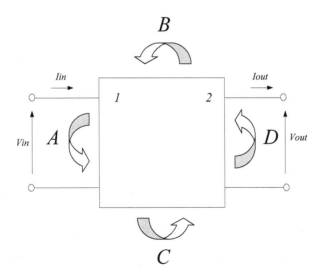

Figure 4.16 Two port system or qudrupole oriented to a transmission line in phasor domain.

Thus, the currents and voltages on both sides of this generic system are related by equations (35)-(37):

$$\begin{bmatrix} Vin \\ Iin \end{bmatrix} = \begin{bmatrix} A & B \\ C & D \end{bmatrix} \cdot \begin{bmatrix} Vout \\ Iout \end{bmatrix} \tag{35}$$

$$Vin = A \cdot Vr + B \cdot Iout \tag{36}$$

$$Iin = C \cdot Vr + D \cdot Iout \tag{37}$$

Standard models for phasor domain are developed starting from these general equations (36)-(37). Depending on the physical phenomena that are considered (section 4.2), transmission lines can be simplified in one way or in another. The consideration or not of all these physical phenomena depends on the cable length. So, depending on the cable length, there are three general models for overhead transmission lines: short (< 80 km), medium (80 Km to 240 Km) and long (more than 240 Km) [47].

### 4.2.2.2.1   Standard short line models for phasor domain (<80 Km)

The capacitive component of the cable increases with its length (equation (23)), so for short overhead cables, the capacitive component is generally small. In the same way, the admittance G represents a leakage current, which depends on the insulation material's conductivity. Usually this material has a low conductivity and the associated resistivity is very high. Therefore, for short lines with very low capacitive component, the capacitive component and the admittance (a high resistivity in parallel) of the line can be neglected [51].

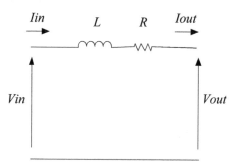

Figure 4.17 Standard cable model for short lengths.

Where: L represents the phenomenon of the self inductance and R is the resistance of the cable.

The ABDC parameters taken into account these simplifications are:

$$A = 1 \; ; \; B = Z \; ; \; C = 0 \; ; \; D = 1 \tag{38}$$

Where: Z is the total series impedance.

If these constants are replaced in the general equation of the standard model (equation (35)) is possible to obtain the general equations of the model for short lines.

### 4.2.2.2.2  Standard medium line models for phasor domain (80Km<length< 240Km)

In cables of medium length, the capacitive component is bigger than in the case before (short lines). So, for lines with this length, the capacitive component has to be considered to obtain an accurate cable model. Therefore, in this case, as the cable model the so called nominal "π" circuit is used, Figure 4.18.

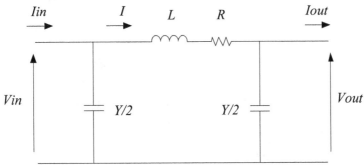

Figure 4.18 Standard cable model for medium cable lengths. Nominal-π circuit.

From this model, it is possible to obtain the following equations to relate Vs, Vr, Ir and Is:

$$Vin = Vout + IZ \tag{39}$$

$$I = Iout + Vout \cdot \frac{Y}{2} \tag{40}$$

$$Iin = Iout + Vin \cdot \frac{Y}{2} \tag{41}$$

Based on these equations, the complex constants A, B, C and D have the following expressions:

$$A = D = 1 + \frac{ZY}{2} \tag{42}$$

$$B = Z \tag{43}$$

$$C = \left(1 + \frac{ZY}{4}\right) \tag{44}$$

Where: Y is the total shunt admittance and Z is the total series impedance.

As in the previous case, if these constants are replaced in the general equation of the standard model (equation (35)) is possible to obtain the general equations of the model for medium lines.

### 4.2.2.2.3 Standard long line models for phasor domain (> 240Km)

In this case, the evolution of the traveling waves and the refraction at the end of the line must be taken into account. So, this model has the general equations shown in (45) - (48).

Usually the conditions of the current and voltage are required at the end of the line (when $x$ = $l$, length of the line). Thus, the general equations of the system are simplified:

$$Vout = Vin \cdot \cosh(\gamma l) - Iin \cdot Zc \cdot \sinh(\gamma l) \tag{45}$$

$$Iout = Iin \cdot \cosh(\gamma l) - \frac{Vin}{Zc} \cdot \sinh(\gamma l) \tag{46}$$

$$Vin = Vout \cdot \cosh(\gamma l) - Iout \cdot Zc \cdot \sinh(\gamma l) \tag{47}$$

$$Iin = \frac{Vout}{Zc} \cdot \sinh(\gamma l) + Iout \cdot \cosh(\gamma l) \tag{48}$$

Where:

$$\cosh(\gamma l) = \left( \frac{e^{\gamma l} + e^{-\gamma l}}{2} \right) \tag{49}$$

$$senh(\gamma l) = \left( \frac{e^{\gamma l} - e^{-\gamma l}}{2} \right) \tag{50}$$

$$Z_c = \sqrt{\frac{L}{C}} \tag{51}$$

In this way, the ABDC constants or the cuadripole constants have the following expressions:

$$A = D = \cosh(\sqrt{ZY}) \tag{52}$$

$$B = \sqrt{\frac{Z}{Y}} \cosh(\sqrt{ZY}) \tag{53}$$

$$C = \sqrt{\frac{Y}{Z}} \cosh(\sqrt{ZY}) \tag{54}$$

However, it is possible to simplify even more those expressions by using equivalent mathematical series instead of the hyperbolic functions:

$$A = D = \cosh\left(\sqrt{ZY}\right) = 1 + \frac{YZ}{2} + \frac{Y^2Z^2}{24} + \frac{Y^3Z^3}{720} + ....$$ (55)

$$B = Z \cdot \left(1 + \frac{YZ}{6} + \frac{Y^2Z^2}{120} + \frac{Y^3Z^3}{540} + ....\right)$$ (56)

$$C = Y \cdot \left(1 + \frac{YZ}{6} + \frac{Y^2Z^2}{120} + \frac{Y^3Z^3}{540} + ....\right)$$ (57)

Usually three terms of the series are enough to model overhead transmission lines with less than 500 km, nevertheless, the first two terms are enough in most cases. Consequently, the equivalent circuit for long transmission lines is an equivalent "п" circuit that can be expressed as follows:

$$\frac{A-1}{B} = \frac{Y}{2}$$ (58)

$$B = Z$$ (59)

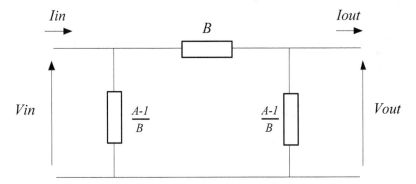

Figure 4.19 Standard cable model for long cable lengths. Equivalent п circuit.

As is distinguished in the beginning of this section, the cable model developed in the present section, depicted in Figure 4.19, only represents the cable for the conditions at the ends of the line. If the results in the intermediate points of the cable are required, the equations (45) - (48) must be considered. Another option is the use of several equivalent п circuits.

#### 4.2.2.2.4 Sequence of single phase "π" segments

If the cable model has to represent the cable in a frequency range bigger than the fundamental frequency (50-60Hz), one option can be the use of several single phase "π" segments in series.

Cascading several identical "π" sections or circuits, as shown in Figure 4.20, it is possible to obtain a lumped parameters model which is an approximation of a distributed parameters model.

Cable models based on distributed parameters have an infinite number of states, but models based on "π" sections have a finite number of states, as many as "π" sections. Therefore, depending on the number of "π" sections used to model the cable, the model is more similar to a distributed parameters model and as a result, the model is able to represent the cable in a bigger frequency range.

Thus, the required number of "π" sections for a specific model depends on the required frequency range and the length of the line. In Figure 4.20 are represented N "π-nominal" circuits in cascade (from now only "π") [52]:

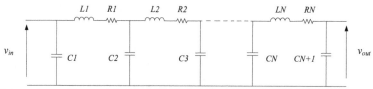

Figure 4.20 Submarine cable modeled as N "π"circuits.

To obtain the relation between the input variables and the output variables, the system is represented as several two-port networks (quadrupoles) with several ABDC parameters Figure 4.21

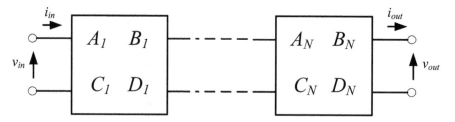

Figure 4.21 N cuadrupoles in series.

It is important to keep in mind that each section is a part of the same cable, thus the output impedance of any cable section and the input impedance of the next section are the same. Consequently, there are not refractions between sections.

Moreover, if all the sections have the same length, as is usual, their associated "π" circuits are the same. The cable model can be represented with equation (60) [50].

$$\begin{bmatrix} v_{in} \\ i_{in} \end{bmatrix} = \begin{bmatrix} A & B \\ C & D \end{bmatrix}^N \cdot \begin{bmatrix} v_{out} \\ i_{out} \end{bmatrix} \tag{60}$$

So, the equivalent ABDC parameters for several "π" circuit is the product of all the matrixes of all two port sections.

The number of "π" circuits in series for a correct representation of the transmission line depends mainly on the required frequency for the analysis [53]. Thus, for a specific maximum frequency, the maximum length that can be represented by each "π" circuit is limited by equation (61).

$$l_{max} \leq \frac{v_e}{5 f_{max}} \tag{61}$$

Where: $v_e$ is the propagation velocity of the traveling wave, $l_{max}$ is the maximum length that can be represented by each "π" section and $f_{max}$ the maximum frequency.

Another way to obtain the number of "π" circuits in series for a correct representation of the transmission line is used in [54]. As [54] a pretty good approximation to determine the number of required "π" circuits is given by the equation (62), which depends on the following parameters.

- The travelling time ($\tau$) or the propagation velocity.
- The maximum required frequency for the simulation model.
- The length of the line ($l$).

$$f_{max} = \frac{N v_e}{8 l} \tag{62}$$

Where: N is the number of required "π" circuits, $v_e$ is the propagation velocity of the traveling wave, $l$ is the length and $f_{max}$ the maximum required frequency.

### 4.2.2.2.5  Bergeron's traveling wave

To perform a accurate analysis of a cable based on constant and distributed parameters, the Bergeron's travelling wave method is used [55], [56] and [57]. This method is based on the way that the travelling wave is propagated and its refraction at the end of the line.

The bergeron's travelling wave model is developed starting from the general equations of transmission lines (29)-(30), but in this case the series resistance and the admittance are neglected. So the transmission line is considered as a lossless line and all the global losses are considered at both ends of the line.

In this way the general equations of the transmission lines can be simplified into:

$$\frac{\partial V}{\partial x} = -L \frac{\partial I}{\partial t} \tag{63}$$

$$\frac{\partial I}{\partial x} = -C \frac{\partial V}{\partial t} \tag{64}$$

Then, if another partial derivative is performed on these equations, the following second degree lineal differential equation is obtained.

$$\frac{\partial^2 I}{\partial x^2} = LC \frac{\partial^2 I}{\partial t^2} \tag{65}$$

Applying D'Alenbert on equation (65), it is possible to obtain the generic expressions for voltage and current:

$$I(x,t) = f_1(x - v_e t) + f_2(x + v_e t) \tag{66}$$

$$V(x,t) = Z_c [f_1(x - v_e t) + f_2(x + v_e t)] \tag{67}$$

Find the particular solution of these equations ((66) and (67)) is very complex. Therefore, for work with them computational models in EMTP (Electro Magnetic transients program) are used. So, the model is adapted to EMTP.

Usually the conditions of the current and voltage are required at the end of the line, i.e. when $x = l$. Thus, this model for EMTP is only valid for results at the end of the lines.

However, if results on an intermediate point of the line are required, it is possible to obtain those by using several Bergeron's travelling wave models in series, i.e. dividing the cable in several sections and modeling each one with traveling wave model. In this case also, due to the fact that those sections are sections of the same cable (the same input impedance), there are not refractions between sections.

The traveling wave does not appear in the other end of the cable instantly. Therefore, the ends of the cable are decoupled with a time constant and there are not any changes in the voltage and current values until expires the period of time necessary for the wave to cross the line ($\tau$). The generic equivalent circuit of the model is shown in Figure 4.22.

$$\tau = \frac{l}{v_e} = l\sqrt{LC} \tag{68}$$

The general equations of the model are:

$$e_k(t) + Z_c i_{km} = e_m(t - \tau) + Z_c(-i_{mk}(t + \tau)) \tag{69}$$

Where the current sources represents the time delay:

$$I_k(t - \tau) = -\frac{1}{Z_c}(e_m(t - \tau)) - i_{mk}(t + \tau) \tag{70}$$

$$I_m(t - \tau) = -\frac{1}{Z_c}(e_k(t - \tau)) - i_{km}(t + \tau) \tag{71}$$

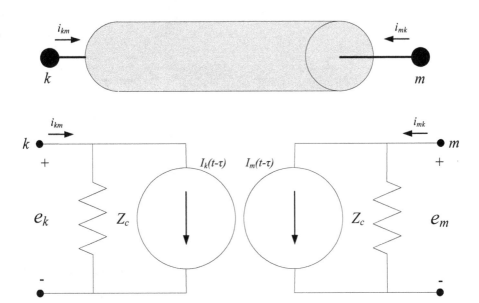

Figure 4.22 Equivalent model for Bergeron's travelling wave.

### 4.2.2.3 Frequency dependent models

If a very accurate analysis is required, the variation of the parameters with the frequency variation must be taken into account. Due to the fact that, the skin effect of the transmission lines affects significantly to the system resonances [58].

The resistive component of a transmission line is only a little part of the total impedance when the system is not in resonance. However, this series resistance is very important when the system is in resonance. If the circuit is in resonance, the imaginary components of the impedance are balanced, so the resistive component determines the total impedance.

Therefore, if the required frequency range for the model is wide, several times bigger than the fundamental frequency (the frequency used to estimate the electrical parameters of the model), these variations have to be taken into account, to obtain accurate results,

### 4.2.2.3.1 Frequency dependent model in modal domain (J. Marti)

One way to obtain more accurate models is developing the transmission line equations in frequency domain considering distributed parameters, due to the distributed nature of the losses and frequency dependent parameters.

In these two fundamentals is developed the model of J. Marti [50] and [59].The equivalent circuit of this model is based in voltage sources, not like Bergeron's model which is based on current sources, as can be seen in Figure 4.23. Another difference with the Bergeron's model are the losses, in this case, the losses are represented by impedances in series.

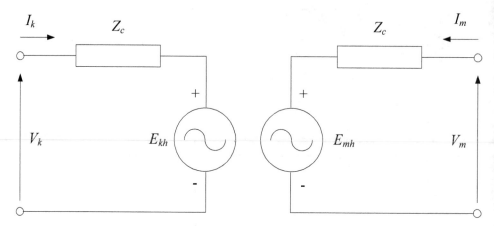

Figure 4.23 Equivalent circuit for J. Marti model in frequency domain.

The general equations of this model, i.e. the equations used to describe the behaviour of the system are (72) and (73):

$$V_k(\omega) = Z_c(\omega) \cdot I_k(\omega) + E_{mh}(\omega) \tag{72}$$

$$V_m(\omega) = Z_c(\omega) \cdot I_m(\omega) + E_{kh}(\omega) \tag{73}$$

Where:

$$E_{mh}(\omega) = A(\omega) \cdot F_{pk} = [V_k(\omega) + Z_c(\omega)I_k(\omega)]e^{-\gamma\omega l} \tag{74}$$

$$E_{kh}(\omega) = A(\omega) \cdot F_{pm} = [V_m(\omega) + Z_c(\omega)I_m(\omega)]e^{-\gamma\omega l} \tag{75}$$

This model is frequency dependent, in both, in the characteristic impedance (equation (76)) and in propagation constant (equation (77)).

$$Z_c(\omega) = \sqrt{\frac{R'(\omega) + i\omega L'(\omega)}{G'(\omega) + i\omega C'(\omega)}} \tag{76}$$

$$\gamma(\omega) = \sqrt{(R'(\omega) + i\omega L'(\omega))(G'(\omega) + i\omega C'(\omega))} \tag{77}$$

The J. Marti model it is not very accurate at low frequencies and for very short lines, due to the imperfections of the system in time domain [60].

### 4.2.2.3.2 Frequency dependent model in phase domain (Idempotent model)

As a example of this kind of models, the Idempotent model is analyzed [50], [61] and [62]. This model and the J. Marti model have similar fundamentals, but this model solves the problem of J. Marti's model with frequency dependent modal transformation matrixes. This problem is solved because the propagation wave is represented in phase domain.

The equivalent circuit of this transmission line model (the idempotent model) is depicted in Figure 4.24.

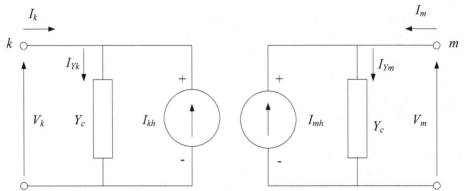

Figure 4.24 Equivalent circuit for the idempotent model.

The general equations of this model are (78) and (79):

$$[Y_c][V_m]-[I_m]=([Y_c][V_k]+[I_k])e^{-[\gamma]l}=[A][F_{pk}] \tag{78}$$

$$[Y_c][V_k]-[I_k]=([Y_c][V_m]+[I_m])e^{-[\gamma]l}=[A][F_{pm}] \tag{79}$$

Where:

$$[I_{mh}]=([Y_c][V_k]+[I_k])[A] \tag{80}$$

$$[I_{kh}]=([Y_c][V_m]+[I_m])[A] \tag{81}$$

The idempotent model with some changes / improvements detailed in [63] is used in PSCAD as the most accurate model. Moreover, the PSCAD user's guide guaranties that its cable model, frequency dependent in phase domain is very accurate [64].

### 4.2.3 Verification of cable models

The complexity and the accuracy are two concerns to be considered to select a cable model. Depending on the required accuracy, the used cable model can be less complex. In some cases, like in cases where the steady state is the analyzed field, the use of complex models does not improve significantly the accuracy of the model. So, there must be a balance between the complexity of the model and required accuracy. In short, depending on the requirements of the analysis, one model or another can be used.

In the present section, to carry out this evaluation of how accurate the considered cable models are and under what conditions can represent the behavior of the transmission line, the PSCAD software is used, software oriented to power electric systems.

PSCAD provides three different cable models with distributed parameters in its standard library: The Bergeron's traveling wave model, the frequency dependent model in modal domain (J. Marti) and frequency dependent model in phase domain [64]. However, for lumped parameter models, like a "π" circuit, there is not any specific cable model. So to represent this kind of models, resistors, capacitors and inductors have to be used.

In the present work, two of the cable models analyzed in the previous section are considered to evaluate their accuracy (Bergeron's travelling wave model and a "π" circuits). In this way, to evaluate their accuracy, the simulation results of the considered two models are compared with the frequency dependent model in phase domain used by PSCAD that has been successfully validated experimentally in [65] and [66].

To work with those distributed parameter models, PSCAD calculates the RGLC parameters for the electric representation of the line by its own, based on physical parameters of the cable.

The basic geometry used by PSCAD to represent the cable is made up by concentric and homogeneous conductor and insulation layers. This approximation of the cable used by PSCAD has even more layers than the example analyzed in section 3.2.1, i.e. presents more conductor and insulation layers. But, a real cable has more layers than the approximation used by PSCAD and even parts that cannot be represented in this approximation. Therefore, to represent correctly a specific cable in PSCAD, some physical parameters of the cable have to be corrected before to fill into the PSCAD template [65].

However, to carry out this comparison, the correction of the physical parameters of the cable does not make any sense, because the physical parameter correction only affects to the electric representation, i.e. to the estimation of the RGLC parameters performed by PSCAD and all the compared models in the present section are based on the same electric representation. Therefore, the parameter correction is explained in section 4.2.4.).

### 4.2.3.1 Comparative of transient response for different cable models
In this section two cable models are compared with the validated model, in order to know how accurate are and in which cases are valid, starting from the simplest (unique "π" circuit) to Bergeron's model (the simplest model considering distributed parameters).

With a sharp voltage variation like a step or pulse, the most frequencies of the spectrum are excited. Thus, it is possible to compare the results and determine if the frequency responses are similar or not. This is the way used by [65] and [66] to determined the PSCAD frequency dependent model in the phase domain as a valid one.

In [65] and [66] to carry out the validation of the cable model, the results obtained by applying a voltage step to a 20kV XLPE cable are compared with the simulation results obtained in PSCAD by applying the same voltage step.

In this section the same procedure is used. Therefore, using the simulation scenario depicted in Figure 4.25, the transient responses for a voltage step for different models are obtained, in order to compare the results and evaluate their similarity.

In this case, instead of a DC voltage step, is used an AC voltage step, i.e. the AC voltage source is suddenly connected to the cable. To carry out this analysis, the breaker is connected in the worst case, when the AC voltage has the maximum value.

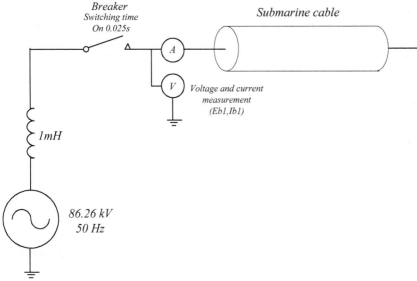

Figure 4.25 Simulation scenario in PSCAD for the verification of cable models.

The simulations are based on a cable of 150kV – 1088 A (courtesy by General Cable) with 50 Km length. The characteristics of this cable are shown in Figure 4.26.

| | Radius | 21.75 mm |
|---|---|---|
| | Conductivity of | (Copper) |
| Conductor | the material | 1.7 e-8 |
| | Magnetic permeability | 1 |
| | thickness | 10 mm |
| | Relative | 2.3 |
| Insulation | permittivity | |
| | Magnetic permeability | 1 |
| | thickness | 0.8 mm |
| | Conductivity of | (Copper) |
| Shield | the material | 1.7 e-8 |
| | Magnetic permeability | 1 |
| | thickness | 42.3 mm |
| | Relative | 2.3 |
| Sheath | permittivity | |
| | Magnetic permeability | 1 |

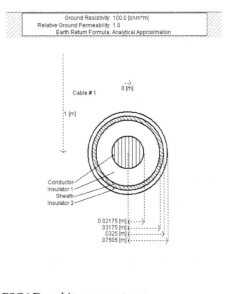

Figure 4.26 Graphic representation of the cable in PSCAD and its parameters.

From the physical characteristics of Figure 4.26, PSCAD solves / estimates the equivalent impedances (RLGC parameters) for the electric representation of the cable (Figure 4.11 for constant parameter models and Figure 4.13 for frequency dependent models). Upon this electric representation, the considered models (described in section 4.2.2.) are developed, unless for the unique "π" circuit model, i.e. to the Bergeon's travelling wave model and frequency dependent model in phase domain.

As is distinguished at the beginning of this section, for lumped parameter models PSCAD does not provides any specific cable model. So, the unique "π" circuit model is defined directly by its RLC parameters. Thus, the cable with the physical parameters described in Figure 4.26 can be defined electrically at 50 Hz with the RLC parameters depicted in Table 4.1.

| 50 Km | Resistivity (ohm) | Inductivity (mH) | Capacitance (uF) |
|---|---|---|---|
| "π" Parameters (50 Hz) | 6.09 | 8.8 | 16.92 |

Table 4.1 RLC parameters for the equivalent "π" circuit which models the 50Km cable for 50Hz.

### 4.2.3.1.1  Comparative of transient responses for the "π" circuit model

Firstly, the comparison of the unique "π" circuit model and the PSCAD frequency dependent model in phase domain is carried out. To this end, the procedure described in the previous section 4.2.3.1 is used.

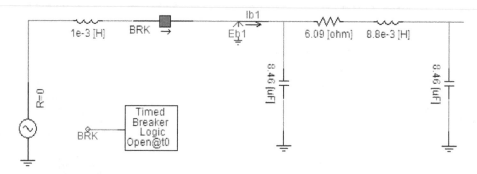

Figure 4.27 Simulation scenario in PSCAD for the "π" circuit.

The simulation results of the submarine cable modeled as a unique "π" circuit and the validated frequency dependent model in phase domain upon a AC step are shown in Figure 4.28 and Figure 4.29. It is easy to observe on these results, that only one "π" circuit is not enough to represent adequately the transient response of the transmission line. The current peak immediately subsequent to the close of the breaker is more than the double compared to the validated model. Moreover, the oscillation frequencies and the duration of the transients are substantially different.

However, it is important highlight that the mean values after the transient of the two models are very similar. So, despite its simplicity, for the steady state, this model can be a reasonably good approximation.

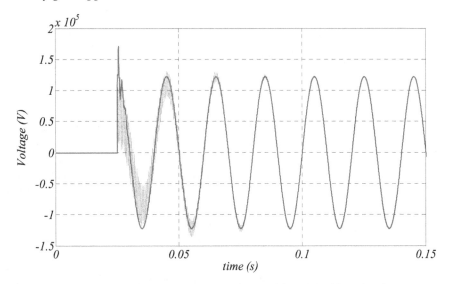

Figure 4.28 Comparison of the input voltages of the cable (Eb1): (blue) for the "π" circuit and (red) for the frequency-dependent model in phase domain of PSCAD.

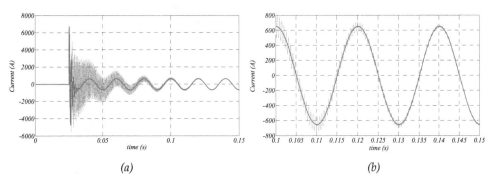

Figure 4.29 Comparison of the input current of the cable (Ib1): (blue) for the "π" circuit and (red) for the frequency-dependent model in phase domain of PSCAD.

### 4.2.3.1.2  Comparative of transients for the Bergeron's travelling wave model

The next model to analyze is Bergeron's travelling wave model. The Bergeron's travelling wave model is based on a distributed parameters representation of the transmission line.

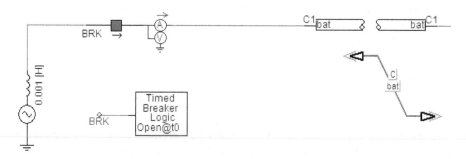

Figure 4.30 Simulation scenario in PSCAD for the Bergeron's travelling wave model.

The simulation results of the scenario depicted in Figure 4.30 are shown in Figure 4.31 and Figure 4.32. In this second case, the transient responses of those models are more similar than in the first case. The first voltage peak, when the breaker is activated, is only a little bit smaller with this model. In the same way, the oscillation frequencies of the transient responses are similar.

With regards to the current, it is possible to see the same behavior pattern. The transient response is a little bit larger in time with Bergeron's travelling wave model and also has more high frequencies.

One reason for this could be the way that the electric parameters (RGLC) are calculated. Bergeron's model is based in an electric representation with constant parameters, so for frequencies far away for the frequency used to calculate those parameters, because of the skin effect (see 4.2.1.1), they cannot represent well the cable. The skin effect increases the resistive component of the cable depending on the frequency and as a result, causes the attenuation of high frequencies.

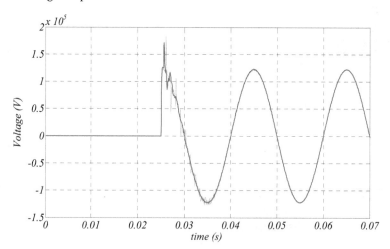

Figure 4.31 Comparison of the input voltages of the cable (Eb1): (blue) for the Bergeron's travelling wave model and (red) for the frequency-dependent model in phase domain of PSCAD.

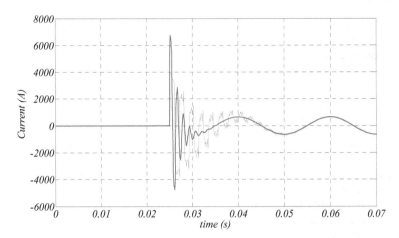

Figure 4.32 Comparison of the input current of the cable (Ib1): (blue) for the Bergeron's travelling wave model and (red) for the frequency-dependent model in phase domain of PSCAD.

In conclusion, this model is accurate for frequencies close to 50Hz (or 60Hz), the frequency usually used to calculate the electric parameters, i.e. it can represent the steady state and low frequencies. Starting from its nominal frequency, if the bigger is the frequency variation, the bigger is the deviation in the frequency response of the system (less accuracy).

### 4.2.4 Cable parameter adaptation for PSCAD

Based on the physical characteristics of Figure 4.26, PSCAD solves / estimates the equivalent impedances (RLGC parameters) for the electric representation of the cable. Thus, for complex models, where are required a lot of parameters and detailed electric specifications, the definition of the cable is more simple. However, this way has a drawback, as is explained in section 4.2.3, the template provided by PSCAD is an approximation and it cannot represent complex cable structures.

The PSCAD template has concentric, circular and homogeneous layers to introduce the data of the cable. Even though, some subsea cables are made up with other physic characteristics, such as: semiconductor layers, conductors made up by a crown of strands, the fill between conductors or other non concentric elements.

Due to the impossibility to fill in correctly the data of the cable to the PSCAD software, the physic parameters of the cable have to be corrected. The purpose of this correction is to achieve the same value for the electric parameters estimated by PSCAD and the parameters measured by the cable manufacturer. Therefore, some parameters of the conductor, shield and insulation are corrected.

The physic characteristics of the cable provided by the manufacturer (Courtesy by General Cable) are shown in Table 4.2

| Parameter | Value |
|---|---|
| Rated voltage | 87 / 150kV |
| Rated current | 1088 A |
| Conductors cross section: | 1.200 mm² |
| Separation between conductors: | 97.839996 mm |
| Buried depth | 1 m |
| Shields cross section | 30 mm² |
| Shield type: | Metallic strip |
| Armor type: | Strands crown |
| Diameter of conductor | 43,5 mm |
| Insulation thickness | 20 mm |
| Diameter upon the insulation | 88,5 mm |
| Diameter down the sheath: | 215,6 mm |
| Diameter down the armor: | 226,7 mm |
| Sheath thickness: | 8,9 mm |
| External diameter: | 244,5 mm |
| Relative dielectric constant: | 2,50 |
| Resistivity of the conductor d.c. at 20°C: | 0,0151 Ohm/km |
| Resistivity of the conductor a.c. | 0,0205 Ohm/km |
| Resistivity of the shield d.c. at 20°C: | 0,6264 Ohm/km |
| Nominal capacitance of the cable: | 0,233 µF/km |
| Inductance of the cable: | 0,352 mH/km |

Table 4.2 Cable characteristics provided by General Cable.

#### 4.2.4.1 Conductor

Looking at Table 4.2, the conductor has a 43.5 mm diameter and also an effective cross section of 1200 mm². If the conductor is considered as solid core, homogenous and circular (as the template does), the cross section for this diameter (equation (65)) it is not the same.

$$A_c = \pi \cdot r_c^2 = \pi \cdot 21.75^2 = 1486.17 mm^2 \tag{82}$$

If the conductor is considered as solid core, homogenous and circular, the effective cross section is 1486.17 mm². Therefore, to solve this difference it is necessary to correct the resistivity of the conductor $\rho$.

Like first step, the real resistivity of the conductor (based on the data of the cable given by the manufacturer) is calculated, equations (83) - (84).

$$R_{DC} = \frac{\rho_c \cdot l}{A_c} = 0.0151 \ ohm/Km \tag{83}$$

$$\rho_c = \frac{R \cdot A_c}{l} = 1.812 \cdot 10^{-8} \tag{84}$$

Where: $\rho_c$ is the resistivity, $l$ is the length of the cable and $A_c$ is the effective cross section of the conductor (1200 mm²).

Then, the correction is applied to the resistivity of the conductor's material, i.e. the resistivity data to fill into the PSCAD template is changed to maintain the same absolute resistance of the conductor. So, the corrected resistivity depends on the effective cross section and the real cross section [65].

$$\rho_c' = \rho_c \frac{\pi \cdot r_c^2}{A_c} = 2.24412 \cdot 10^{-8} \tag{85}$$

To verify this estimation, the absolute resistance of the conductor in two cases is calculated, for 50 Hz and for direct current. In this way, these results can be compared with the characteristics provided by the manufacturer (Table 4.2).

$$R_{a.c.}(50) = \frac{\rho_c}{\delta_{50}} \cdot \frac{l}{\pi(D-\delta_{50})} = \frac{2.244 \cdot 10^{-8}}{0.010662} \cdot \frac{1000}{\pi(0.0435-0.010662)} = 0.0204 \ ohm/Km \tag{86}$$

$$\delta_{50} = \sqrt{\frac{2 \cdot \rho_c}{\omega \cdot \mu}} = \sqrt{\frac{2 \cdot 2.244 \cdot 10^{-8}}{2 \cdot \pi \cdot 50 \cdot 4 \cdot \pi \cdot 10^{-7}}} = 0.010662 \tag{87}$$

Where: $l$ is the length of the cable, $D$ is the diameter of the conductor, $\rho_c$ is the resistivity, $\omega$ is the angular speed of the current ($2\pi f$), $\mu$ is the absolute magnetic permeability of the conductor ($\mu_0 \mu_r$), $\mu_0$ is the magnetic constant or the permeability of the free space ( $4\pi \times 10^{-7}$ N/A$^2$) and $\mu_r$ is the relative magnetic permeability.

Comparing the results obtained in equations (86) with the data given by the manufacturer (Table 4.2.), it is possible to observe practically the same values. Consequently, the correction performed to the resistivity of the conductor is reasonable accurate.

### 4.2.4.2 Shield

The shield is something more than only a metallic strip, the shield can be made up with multiple layers. Looking at Table 4.2, the conductor has a 43.5 mm diameter and the insulation has a thickness of 20 mm. However, the diameter upon the insulation is 88.5mm. So, there is an undefined layer of 2.5 mm.

Considering from the data provided by the manufacturer that the shield has only 30mm$^2$ of cross section. It is possible to deduce that one of them: the shield or the insulation has more complex structure than the PSCAD template.

Therefore, in order to maintain the shield with a 30 mm$^2$ thickness and the same capacitive component for the cable, the diameter of the insulation and its relative permeability has to be changed.

Assuming that the outer diameter of the shield's conductor layer is 88.5 mm, it is possible to obtain the inner diameter of the shield, equations (88) - (90).

$$A_s = R_s^2 - r_s^2 \tag{88}$$

$$30mm2 = 44.45^2 - r_s^2 \tag{89}$$

$$r_s = \sqrt{44.25^2 - 30} = 43.9mm \tag{90}$$

In the present analysis the Shield of the cable is metal strip kind. However, there are also other types of shields, like strands crown kind. In these cases it is necessary to carry out another correction [65].

### 4.2.4.3 Insulation

To correct the area of the shield, the radius of the insulation is modified. So, the value of the capacitive component using the radius calculated in the equation (90) is not the same of the characteristic provided by the manufactures, equation (91).

$$C = \frac{\varepsilon_r}{17.97 \cdot \ln\left(\frac{b}{a}\right)} = \frac{2.5}{17.97 \cdot \ln\left(\frac{43.9}{21.75}\right)} = 0.198 \;\; \mu F / Km \tag{91}$$

Therefore, to represent correctly in PSCAD the capacitive component of the submarine cable, the dielectric constant has to be corrected, equations (92) - (93).

$$\varepsilon_r = 0.233 \cdot 17.97 \cdot \ln\left(\frac{43.9}{21.75}\right) = 2.94 \tag{92}$$

$$C = \frac{\varepsilon_r}{17.97 \cdot \ln\left(\frac{b}{a}\right)} = \frac{2.94}{17.97 \cdot \ln\left(\frac{43.9}{21.75}\right)} = 0.233 \;\; \mu F / Km \tag{93}$$

### 4.2.4.4 Measure with PSCAD the adapted parameters

To validate the parameters correction carried out in the preceding sections, a equivalent submarine cable in PSCAD (Figure 4.33) based on the physic data of the cable shown in Table 4.3. is defined. In this way, it is possible to obtain the electric parameters via PSCAD to compare them with the electric data provided by the manufacturer.

Table shows the parameters of the transmission line solved by PSCAD based on the physical parameters of Figure 4.33.

| | Resistivity Seq + | Inductivity Seq + | Capacitance Seq + |
|---|---|---|---|
| Electric parameters (50 Hz) | 0.0311* Ohm/km | 0.334 mH/km | 0.233 µF/km |

*Resistivity without taking into account the shield, conductor only 0.0190 Ohm/km

Table 4.3 RGLC electrical parameters calculated by PSCAD based on the physical dimensions and characteristics.

As a conclusion, the electrical parameters calculated by PSCAD (Table 4.3) in comparison with the parameters specified by the manufacturer are substantially similar.

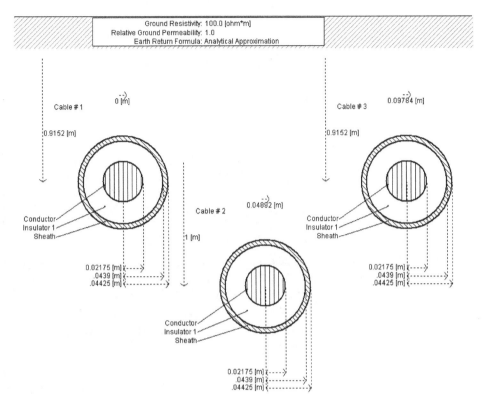

Figure 4.33 Graphic representation in PSCAD of the three-phase cable.

## 4.3 Reactive power Management in subsea power cables

### 4.3.1 Introduction

For the construction of the onshore high voltage transmission lines, almost exclusively air insulated conductors are used. However, the submarine cables are placed in very different environment which requires other kind of insulation.

Therefore, due to their construction characteristics (conductor, insulation, shield and armor), the submarine cables have a high capacitive component (see section 4.2). As a consequence, the transmission of the energy through these cables in AC provokes charge / discharge currents. This current causes a reactive power.

The maximum current capable to transfer a cable is determined by its construction characteristics (cross section of the conductor and its thermal characteristics). Thus, the active current capable to carry a cable is limited by charge / discharge current (reactive current) flowing through the cable (equation (94)).

$$|I| = \sqrt{I_{active}^2 + I_{reactive}^2} \qquad (94)$$

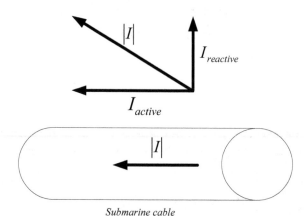

*Submarine cable*

Figure 4.34 Graphic representation of the current through the cable.

Moreover, the capacitive component of the cable increases with the length of the cable (equation (23)). So, there is a specific length for each voltage level where the reactive current generated in the cable is the same of the rated current. This limit depends on the cable voltage and capacitive component (length).

Considering the transmission frequency constant (50Hz-60Hz), the expression of the reactive current is shown in equations (95) - (97).

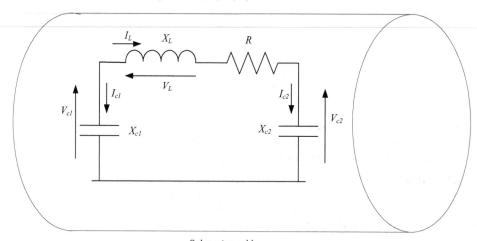

*Submarine cable*

Figure 4.35 Simplified phase to neutral point (not physical) representation of a mono-phase submarine cable using a "п" model.

$$I_{c1} = \frac{|V_{c1}|}{|X_{c1}|} \approx I_{c2} = \frac{|V_{c2}|}{|X_{c2}|} \tag{95}$$

$$\left|X_{c1}\right| = \left|X_{c2}\right| = \frac{1}{\omega C_2} = \frac{1}{\omega C_1} \tag{96}$$

$$I_{c1} = \left|V_{c1}\right| \cdot \omega \cdot C_1 \approx I_{c2} = \left|V_{c2}\right| \cdot \omega \cdot C_2 \tag{97}$$

Where: $I_{C1}$ and $I_{C2}$ are the reactive currents, $|V_{C1}|$ and $|V_{C2}|$ are the magnitude of the applied voltage, $\omega$ is the pulsation and $C_1$ and $C_2$ represents the capacitive component of the cable.

So, if the reactive current reaches the rated current of the cable, this cable cannot transfer any active power.

To understand better this problem, the reactive power generated in the cable (caused by reactive current) for three examples is calculated, Table 4.5 Examples with different voltages and rated powers, the electrical characteristics of the cables used in those examples are depicted in Table 4.4

| Cable | Vn (kV) | In(A) | Rac($\Omega$/Km) | L (mH/Km) | C ($\mu$F/Km) |
|-------|---------|-------|---------|-----------|------------|
| A | 36 | 911 | 0.0341 | 0.294 | 0.331 |
| B | 150 | 1088 | 0.0205 | 0.352 | 0.233 |
| C | 220 | 1055 | 0.048 | 0.37 | 0.18 |

Table 4.4 Characteristics of the submarine cables.

| Cable | Vn (kV-50Hz) | Reactive current (A/Km) | Reactive power (MVAR/Km) |
|-------|--------------|-------------------------|--------------------------|
| A | 36 | 2.2 | 0.134 |
| B | 150 | 6.33 | 1.6455 |
| C | 220 | 7.18 | 2.736 |

Table 4.5 Reactive power generated in the submarine cables.

Therefore, to make more efficient the power transmission system, it is necessary to minimize the reactive current flowing through the cable by managing the reactive current. This management of the reactive power has to be oriented to increase the active power transmission capability of the transmission lines and to reduce the conduction losses.

### 4.3.2 Types of reactive power management
With low power factors, the conduction losses in the cable are higher and the required conductor size is higher. Therefore, most of the grid codes for wind farms have as a requirement a power factor close to unit at the point of common coupling (see Appendix B: Power Factor Requirements at the Point of Common Coupling). Consequently, the reactive power compensation is necessary at least at the PCC.

If the reactive power flowing through the transmission system is managed in addition to the compensation at the PCC, makes the transmission system more efficient. With regards to the reactive power management classification, these can be classified by two parameters:
- The location of the compensation: at one end only or in both ends.
- The characteristics of the compensation: static or dynamic. The compensation device can compensate always the same reactive power at the same voltage or on the contrary depending on the requirements.

### 4.3.2.1 Reactive power compensation at one end or in both ends

In the first case, the reactive power compensation at one end is evaluated. This reactive power compensation is made by injecting reactive current in one end only, at the onshore side. So, there is not a reactive power management of the transmission line.

Basically, the energy is generated with a unitary power factor. This energy is transmitted through the submarine cables to the onshore substation, where the reactive power generated in the cable is compensated to integrate the energy in the main grid, Figure 4.36 (a).

The best option to minimize the reactive current as much as possible is the compensation of the reactive power generated in the cable along all its length. This option, allows the use of the cable without length limit and the reduction of the conduction losses to the minimum. But this option is not easy to carry out, due to the fact that the cables are placed in the seabed.

Therefore, if it is not possible to place reactive power compensation at intermediate points of the cable, the next best option is the compensation of the reactive power of the cable at both ends [9].

In this kind of reactive power management, an inductive reactive current is injected at the offshore end of the submarine cable, thus, due to the capacitive and distributed nature of the submarine cable, this inductive reactive current is neutralized along the length of the submarine cable.

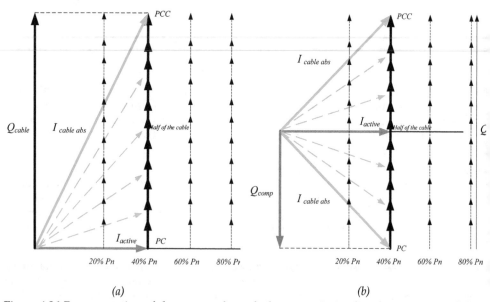

(a)            (b)

Figure 4.36 Representation of the current through the transmission line for two ways of the reactive power management: (a) reactive power compensation at one end and (b) reactive power compensation in both ends.

So, if this injected inductive current is exactly the half of the total capacitive reactive current generated by the cable, the inductive current is neutralized along the cable length until the middle of the cable. At this point, the inductive reactive current injected in the offshore end is completely neutralized and the power factor is unitary. Consequently, at the other end of the cable only appears the reactive power generated by the half of the cable, i.e. only the half amount of the capacitive reactive power generated in the cable, which is compensated by the compensation device at this end (onshore).

In this way, it is possible to minimize the maximum current flowing through the cable as well as the conductive losses, Figure 4.36 (b).

To compare these two different ways to manage the reactive power, an example is defined. This example is based on the cable "B" (Table 4.4) of 50 Km. With regards to the cable model, several "π" circuits in series are used (see section 4.2.2.2.4).

The analysis uses this cable model, because despite its simplicity, it is possible to obtain accurate results for the steady state and allows the measure of the results at intermediate points of the cable. More specifically, one "π" circuit for each kilometer is used.

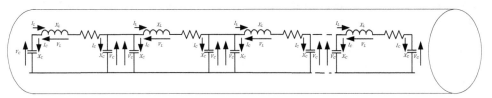

*Submarine cable*

Figure 4.37 Simplified phase to neutral point (not physical) representation of a mono-phase submarine cable using N "π" circuits in series.

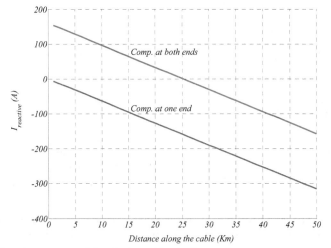

Figure 4.38 Reactive current through the cable, (red) reactive power compensation at both ends and (blue) reactive power compensation at one end only. For the cable B (150kV) and 50 Km.

Therefore, as can be seen at Figure 4.38, using the reactive power management which compensate the reactive power at both ends, the maximum current value decreases in comparison with the management way which injects the reactive power in one end only. As a result, the capability of the cable to carry active power increases. In the same way, the conduction losses are reduced.

Going more depth into the analysis, a more detailed comparison is carried out. In this second case, more cable lengths, different voltages and several transmitted active power levels are considered.

For that purpose, two cables are considered: the cables "A" and "B" of Table 4.4. As in the case before, there are modeled like several "π" circuits in series.

Based on these two cables, the total current along the line is obtained, for these two ways of the reactive management. In this case, two main scenarios are considered: 30 MW of transmitted active power with 36 kV of transmission voltage and 150 MW of transmitted active power with 150 kV of transmission voltage.

In each one of those scenarios, four cable lengths are considered: 50Km, 100Km, 150Km and 200Km. The results of the total current along the line for these configurations are depicted in Figure 4.39.

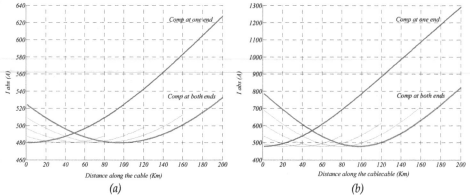

(a)                                                              (b)

Figure 4.39 Total current along the submarine cable depending on cable length, compensation at both ends (red) and onshore compensation only (blue). (a) 30MW-36kV configuration and (b) 150MW-150kV.

Figure 4.39 shows the current along the cable for different lengths. For the case of the only onshore compensation, for different cable lengths, circulates the same current at the same length. On the contrary, with a compensation of 50% of the reactive power at each ends, the minimum current appears in the middle of the cable and the maximum currents at both ends.

This reduction and current distribution pattern along the cable happens for all the cable lengths and all the configurations. The different configurations and the different cable properties only affects on the amount of the generated reactive power, not in this maximum current reduction or in the shape of the current along the cable.

The management of the reactive power through the cable is determinant for long distances, because without the proper reactive power compensation, the cable may not be capable to transmit the required energy.

For example, the scenario with 150MW-150kV (using the cable "B" of Table 4.4) has a rated current of 1088A. But, as can be seen in Figure 4.39, for cable lengths longer than 150Km the required total current is bigger. Therefore, if a reactive power compensation at both ends is not performed, the cable cannot be capable to transmit the required active current.

### 4.3.2.2 Reactive power compensation: fixed or variable

Before to perform the analysis about the compensation characteristic fixed or variable, it is important to know how is generated the reactive power in the cable in more detail.

In addition to the capacitive component; the cable also has an inductive component. So, the reactive power generated in the cable is not constant at a specific voltage, depends on the transmitted power.

The capacitive reactive power generated in a specific cable (specific capacitive component) is determined by the applied voltage; due to this component is a "shunt" impedance. However, the inductive reactive power generated by the cable depends on the amount of active current flowing through the cable (transmitted active power), Figure 4.35. As a consequence, the amount of the reactive power generated by the cable varies.

$$Q_C = \frac{|V_C|^2}{|X_C|} \tag{98}$$

$$Q_L = |I_L|^2 \cdot |X_L| \tag{99}$$

Where: $Q_C$ is the capacitive reactive power per phase, $|V_C|$ is the module of the voltage applied to the capacitive component, $Q_L$ is the inductive reactive power per phase, $|I_L|$ is the module of the current through the inductive component, $|X_L|$ is the module of the inductive impedance and $|X_C|$ is the module of the capacitive impedance.

The philosophy of the reactive power management is the minimization of the total current flowing through the cable by reducing the reactive current flowing through the cable as much as possible.

Thus, to reduce to the minimum the reactive current flowing through the cable by the injection of the inductive current at both ends, it is necessary to inject exactly the half of the capacitive reactive current generated by the cable.

Therefore, because of the variation in the reactive current generated by the cable, the half of this reactive current needed for the optimum reactive power management varies too. This variation at each end is exactly the half of the total variation of the cable.

However, considering the total capacitive reactive current generated by the cable, this variation is quite little. So, the use of big inductances to perform the reactive power compensation is an economic option.

Regarding to the characteristics of those inductances, their inductivity depends on the reactive power generated by the submarine cable: The capacitive reactive power (equation (98)) minus the inductive reactive power (equation (99)), this last one, depending on the transmitted active power. So, the inductivity of those inductances has to be adjusted for a specific transmitted active power.

The best option, the option which reduces to the minimum the required maximum current of the cable, can be achieved adjusting the inductivity for the worst case. The case when the line is transmitting the rated active power. When the cable is transmitting the rated power, the maximum active current is flowing through the cable, so, if the reactive current flowing through the cable is optimized (reduced to the minimum) for this case, the required rated current of the cables is reduced.

The expression of the generated inductive reactive current / power at these inductances is shown in equations (100) and (101).

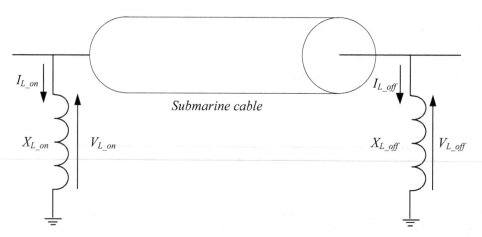

Figure 4.40 Graphic representation of a submarine cable (mono-phase) with inductances at both ends.

$$Q_{Lcomp\_on} = \frac{\left|V_{L\_on}\right|^2}{\left|X_{L\_on}\right|} = \frac{\left|V_{L\_on}\right|^2}{\omega \cdot L\_on} \tag{100}$$

$$L\_on = \frac{\left|V_{L\_on}\right|^2}{\omega \cdot Q_{Lcomp\_on}} \tag{101}$$

Where: $Q_{Lcomp}$ is the module of the inductive reactive power and $|V_L|$ is the module of the applied voltage in the inductance.

### 4.3.3 Comparative of different types of reactive power compensation for a specific scenario in PSCAD

Any change in the power characteristics at the onshore substation, affects only to the circuit which is after this point, i.e. any change in the power factor at this point (PCC), only affects to the characteristics of the power injected in the main grid.

Nevertheless, controlling the power factor at the collector point of the offshore wind farm (the offshore end of the transmission cable), it is possible to control the active and reactive current relation through the transmission line. Due to the fact that this point is the energy "emitter" point. Thus, to carry out the management of the reactive power flowing through the transmission line, the main point to control the reactive power is the collector point.

Therefore, to perform the management of the reactive power flowing through the transmission line, there are three different options:

- Option 1: Without a reactive power management through the cable. The energy is generated with a unitary power factor in the wind farm, i.e. P.F. $\approx 1$ in the "emitter" end of the cable and then transmitted through the submarine cables to the onshore substation. At this point, the reactive power generated in the cable is compensated to integrate the energy into the main grid.
- Option 2: With a rough reactive power management. The inductive reactive power is injected at both ends, but via inductances, always the same quantity for the same voltage.
- Option 3: With an adjusted reactive power management. The inductive reactive power is adjusted dynamically to obtain the same current module at both ends, i.e. the reactive power is injected depending on transmitted active power.

In the present section, a comparative of these three different options is carried out. For this purpose, a scenario using the cable model validated in the previous section is developed. This scenario has a 150 MW rated power, with a transmission voltage of 150 kV and 50 Km submarine cable length.

The objective is to obtain the main parameters of the transmission system upon this scenario for each one of those three management options. The parameters taken into account are: Active power losses, the reactive power generated in the transmission system and the voltage drop.

The analysis is focused exclusively in the transmission line and in the steady state, so, the wind farm is considered as a controlled P,Q source (see Figure 4.41).

The reactive power generated by the submarine cables varies with the amount of the transmitted active power, thus, a range of the transmitted active power is defined. This range is calculated, considering that the wind turbines generate energy with wind speeds between 3.5-30m/s. If the wind farm has 30 wind turbines of 5 MW, depending on the equation (102) explained in section 2.2.3, the range of the generated active power of a wind farm is approximately 5MW to 150 MW.

$$P_t(v) = \frac{1}{2} \cdot \rho \cdot \pi \cdot R^2 \cdot V_{wind}^3 \cdot Cp \qquad (102)$$

Where: $P_t(v)$ is the output power depending on the wind in Watts, ρ is the density of air ( 1.225 measured in kg/m $^3$ at average atmospheric pressure at sea level at 15° C), $r$ = the radius of the rotor measured in meters (63m), $V_{wind}$ = the velocity of the wind measured in m/s and $Cp$ = the power coefficient (0.44).

### 4.3.3.1 Option 1: Without reactive power management (reactive power compensation at one end)

In this first case, the first of the three considered reactive power management options is analyzed. To this end, the scenario illustrated in Figure 4.41 is simulated.

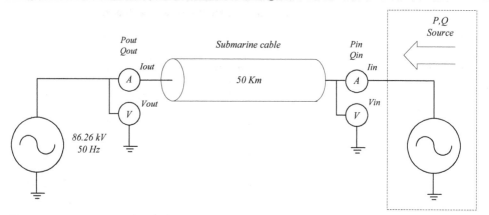

Figure 4.41 Diagram of the first simulation scenario, submarine cable without reactive power management.

In Table 4.6, the simulation results of the transmission system depicted in Figure 4.41 are summarized. These results are obtained in the active power range of the offshore wind farm (5-150MW).

| Pin MW (FP=1) | Generated Q (Qout-Qin) | P losses (Pin-Pout) | ΔV (kV) (Vout-Vin) | Iin | Iout |
|---|---|---|---|---|---|
| 5 | 83.07 MVAR | 0.237 MW | 1.22 | 21 A | 318 A |
| 15 | 83.07 MVAR | 0.251 MW | 1.31 | 58 A | 323 A |
| 25 | 83.05 MVAR | 0.284 MW | 1.425 | 99 A | 333 A |
| 40 | 82.9 MVAR | 0.344 MW | 1.5 | 153 A | 357 A |
| 60 | 82.55 MVAR | 0.488 MW | 1.7 | 234 A | 392 A |
| 80 | 82.01 MVAR | 0.683 MW | 1.87 | 312 A | 440 A |
| 100 | 81.32 MVAR | 0.915 MW | 2.03 | 385 A | 495 A |
| 120 | 80.4 MVAR | 1.22 MW | 2.18 | 459 A | 555 A |
| 150 | 78.75 MVAR | 1.76 MW | 2.38 | 575 A | 648 A |

Table 4.6 Simulation results for the first scenario. Reactive power generated in the cable, active power losses and voltage drop.

As can be seen in Table 4.6, the capacitive reactive power generated in the transmission line is about 83 MVAR. This value is in concordance with the estimation in a simply way of the equation (103) explained in section 4.3.2.2.

$$Q_C = 3 \cdot \frac{|V_C|^2}{|X_C|} = \frac{86.6^2}{273.22} = 82.3 MVAR \tag{103}$$

Where: $Q_C$ is the capacitive reactive power for the three phases, $|V_C|$ is the module of the applied voltage to the capacitive component and $|X_C|$ is the module of the capacitive impedance.

### 4.3.3.2 Option 2: Transmission system with fixed reactive power compensation (at both ends)
In the second case, the management of the reactive power flowing through the transmission line is made via inductive impedances at both ends of the line, (Figure 4.42). The value of the inductance is calculated to compensate exactly the reactive power generated in the cable when this is carrying the rated active power, equations (100) and (101).

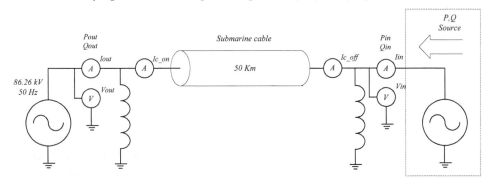

Figure 4.42 Diagram of the second simulation scenario, submarine cable with fixed inductances at both ends.

The results, obtained by the simulation of the defined scenario with the second option for the reactive power management, are depicted in Table 4.7.

Comparing the results in Table 4.6 with the results on Table 4.7, can be seen how the reactive power management reduces significantly the voltage drop in the line. In the same way, with this kind of reactive power management, the active power losses have an important decrement. Especially, in cases when the transmitted active power through the line is less than the 50% of the rated power.

In the considered range of the active power generated for the offshore wind farm (5-150MW), the variation of the reactive power generated in the line is about 4.8 MVAR, i.e. the submarine cables in combination with the inductances have a 4.8 MVAR variation. This value is significantly low in comparison with the reactive power generated in the submarine cables 83 MVAR, approximately the 5%.

| Pin MW (FP=1) | Generated Q (Qout-Qin) | P losses (Pin-Pout) | ΔV (kV) (Vout-Vin) | Iin | Ic_off | Ic_on | Iout |
|---|---|---|---|---|---|---|---|
| 5 | 4.92 MVAR | 0.12 MW | 0.106 | 17 A | 151 A | 171 A | 26 A |
| 15 | 4.88 MVAR | 0.13 MW | 0.19 | 55 A | 162 A | 174 A | 60 A |
| 25 | 4.8 MVAR | 0.16 MW | 0.27 | 95 A | 178 A | 193 A | 99 A |
| 40 | 4.58 MVAR | 0.22 MW | 0.41 | 151 A | 217 A | 230 A | 153 A |
| 60 | 4.12 MVAR | 0.365 MW | 0.596 | 234 A | 274 A | 280 A | 238 A |
| 80 | 3.46 MVAR | 0.56 MW | 0.75 | 310 A | 342 A | 351 A | 314 A |
| 100 | 2.67 MVAR | 0.79 MW | 0.89 | 380 A | 411 A | 416 A | 382 A |
| 120 | 1.66 MVAR | 1.1 MW | 1.03 | 458 A | 482 A | 485 A | 460 A |
| 150 | 0.16 MVAR | 1.65 MW | 1.24 | 572 A | 590 A | 591 A | 572 A |

Table 4.7 Simulation results for the second scenario. Reactive power generated in the transmission system (cable + inductances), active power losses and voltage drop.

This value, the reactive power variation of the cables depending on the transmitted power, is in concordance with the estimation in a simply way of the equation (104) explained in section 4.3.2.2.

$$Q_L = 3 \cdot I_L^2 \cdot |X_L| = 3 \cdot 577^2 \cdot 5.246 = 5.23 MVAR \qquad (104)$$

$$I_L = \frac{P}{3 \cdot V_{\phi n}} = \frac{150 MW}{3 \cdot 86.6 kV} = 577 A \qquad (105)$$

Where: $Q_L$ is the inductive reactive power for the three phases, $|I_L|$ is the module of the current through the inductive component, $|X_L|$ is the module of the inductive impedance, $P$ is the transmitted power through the transmission line and $V_{\phi n}$ is the rated voltage per phase of the transmission system.

### 4.3.3.3 Option 3: Transmission system with variable reactive power compensation (at both ends)

In the third and last case, to achieve the optimum reactive power management, the inductive reactive current is injected at both ends of the cable depending on the transmitted amount of the active power.

As is estimated in the case before, the variation of the reactive power generated in the line (cable + inductances) is about the 5%. So, in this third case, the effect of this variation in the transmission system parameters is analyzed. For this purpose, the simulation of the scenario shows in Figure 4.43 is carried out.

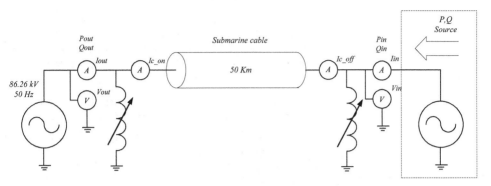

Figure 4.43 Diagram of the third simulation scenario, submarine cable with the injection of the inductive reactive power at both ends depending on the requirements.

The simulation results of the defined scenario (Figure 4.43), for the considered active power range of the offshore wind farm (5-150MW) are summarized in Table 4.8. Notice that in the results of Table 4.8, there are not shown the results of the reactive power generated in the transmission line, because the reactive power of the cable is totally compensated at both ends.

| Pin MW (FP=1) | P losses (Pin-Pout) | ΔV (kV) (Vout-Vin) | Iin= Iout | Ic_on= Ic_off |
|---|---|---|---|---|
| 5 | 0.118 MW | 0.035 | 17 A | 160 A |
| 15 | 0.13 MW | 0.12 | 55 A | 167 A |
| 25 | 0.16 MW | 0.2 | 95 A | 185 A |
| 40 | 0.22 MW | 0.345 | 151 A | 223 A |
| 60 | 0.36 MW | 0.536 | 234 A | 277 A |
| 80 | 0.56 MW | 0.7 | 310 A | 347 A |
| 100 | 0.79 MW | 0.855 | 380 A | 413 A |
| 120 | 1.1 MW | 1.01 | 458 A | 483 A |
| 150 | 1.65 MW | 1.24 | 572 A | 590 A |

Table 4.8 Simulation results for the second scenario. Active power losses and voltage drop.

Comparing the results on Table 4.8, with the results on Table 4.7, it can be seen how the active power losses have not a significantly reduction. With regards to the voltage drop, this has a little reduction only in cases when the transmitted active power through the line is less than the 50% of the rated power.

The fixed inductances at both ends of the cable are fit to achieve the optimum reactive power management with rated active power. So, in cases when the transmitted active power is close to the rated power, with both ways: with fixed inductances and with variable injection of the reactive power, similar results are obtained.

### 4.3.3.4 The effect of the cable length

The reactive power generated in the submarine cables depends on the cable length (Table 4.5), thus, as the length affects to the amount of reactive power to compensate, this aspect has to be analyzed.

In this way, with the increase of the reactive current through the line, the current limit of the cable (1088A in the present case, Table 4.4) has to be taken into account. If the required active current to transmit the rated power of the wind farm is close to the rated current of the cable or the transmission cable is too long, it is possible that without the proper compensation, the cable would not be capable to transmit the required power.

In Table 4.9, the results of the three types of reactive power management (explained in the previous sections) for two cable lengths 50 Km and 150Km are summarized.

| Transmitted active power MW (P.F.=1) | | 50 Km cable length | | | |
|---|---|---|---|---|---|
| | | ΔQ (MVAR) | ΔP (MW) | ΔV (kV) | Imax (A) |
| 5 MW | Option 1 | 83.07 | 0.237 | 1.22 | 318 |
| | Option 2 | 4.92 | 0.12 | 0.106 | 169 |
| | Option 3 | - | 0.118 | 0.035 | 160 |
| 150 MW | Option 1 | 78.75 | 1.76 | 2.38 | 648 |
| | Option 2 | - | 1.65 | 1.24 | 592 |
| | Option 3 | - | 1.65 | 1.24 | 592 |
| | | 150 Km cable length | | | |
| | | ΔQ (MVAR) | ΔP (MW) | ΔV (kV) | Imax (A) |
| 5 MW | Option 1 | 262.4 | 4.73 | 11.47 | 1012 |
| | Option 2 | 11.9 | 1.05 | 0.61 | 504 |
| | Option 3 | - | 1 | 0.075 | 482 |
| 150 MW | Option 1 | 256.9 | 8.4 | 14.6 | **1130** |
| | Option 2 | - | 5.1 | 3.4 | 728 |
| | Option 3 | - | 5.1 | 3.4 | 728 |

Table 4.9 Simulation results for different kind of reactive power management, for two cable lengths 50Km and 150Km.

Notice that without the proper reactive power management, it is impossible to transmit the rated power with the selected cable to 150Km away, because the required current for that purpose is higher than the current limit of the cable.

For cases with a submarine cables of 150Km or longer, providing the transmission system with reactive power management, it is possible to see a higher reduction in the voltage drop and active power losses. Making clear that, the more is the reactive power generated in the cable, the more important is a correct reactive power management.

## 4.4 Chapter conclusions

In this chapter, the main characteristics of the submarine cables are analyzed, such as: their physical structure, the way to represent them electrically or different ways to model it, from the most simply to more complex models. Then, based on a validated cable model several electric aspects of the transmission lines are evaluated.

In this way, from the analysis carried out in this chapter is clear that the transmission of the 150 MW at 150kV to 50Km away is perfectly possible. At least if it is performed with the cable considered in this chapter. Using this cable, the voltage drop of the transmission line is less than 5% and the active power losses are not too high. In the same way, the current limit of the cable is enough to carry the rated active power at any circumstance. So, this scenario is perfectly valid and feasible.

The reactive power management of the submarine cable reduces significantly the active power losses and the voltage drop. This reduction is more obvious, in cases where the transmitted active power is less than the 50% and for long submarine cables (big amount of generated reactive power), i.e. this reduction is more obvious, in cases when the reactive current is high in relation to the active current.

The reactive power management, based on fixed inductances at both ends of the line has similar improvements in comparison with the variable compensation at both ends. Moreover, if those inductances are adjusted for the worst case, using fixed inductances, the maximum voltage drop, the maximum active power losses and the required maximum current for the cable are exactly the same. So, this option is simply and enough for a good reactive power management.

# Chapter 5

# Definition of a Base Scenario

The objective of this book is the analysis of the key issues of the offshore wind farm's energy transmission and grid integration infrastructure by using a representative case. Thus, in the present chapter, a main scenario (base scenario) is defined. The evaluation starts with some generic features of the offshore wind farm, such as: the rated power and the distance to shore. Then, considering these generic features, the main elements of the offshore wind farm are characterized, based on the current state of the technology.

In this way, firstly the number of connections to shore and the transmission voltage level based on the economical optimum are selected (without considering the wind turbines). Thus, based on the developed submarine cable model, on specific location characteristics and specific cost estimations, the transmission cost for three different AC transmission configurations is calculated: single HVAC, various HVAC and MVAC. Then considering this estimated transmission cost, the most cost efficient lay-out is selected for a 150MW wind farm at 50 km to shore.

Once, the main electric connection structure is selected, a base scenario is developed for further analysis. To that end, the offshore wind farm's components are modeled and sized taking special care on wind turbines.

The wind turbines are considered a key issue. So, after the definition of their rated power, the control strategy and the grid side filter, the wind turbines are tested via simulation to verify that the defined wind turbines are suitable to place in an offshore wind farm and if they can fulfill the grid codes.

## 5.1 Wind farm's layout selection

As it has been concluded in chapter 3, to select the proper energy transmission solution, a specific analysis is needed. Therefore, in the present chapter by using a design procedure, based on the location characteristics and a proper reactive power management of the submarine cable of the chapter 4, the most cost efficient energy transmission solution is defined.

In the literature, several analyses about the energy transmission cost based on the produced energy are carried out [67], [68]. These studies are focused on the comparison between AC transmission and DC transmission options. More specifically, the analysis in [68] is based on very high rated powers 400-1000MW. These studies also do not consider in detail the reactive power compensation or different AC transmission options at different rated powers and voltages.

Therefore in this section, using a similar procedure, the cost for different lay-outs focused in AC configurations is estimated. More specifically, considering a wind farm of 150 MW in a location with 9m/s of average wind speed, the transmission cost for several AC lay-outs at different distances to shore is calculated.

### 5.1.1 Considered offshore AC layouts

There are several options to transfer the energy generated in an offshore wind farm to the distribution grid. These options range from the HVDC (High Voltage Direct Current) connection to various MVAC (Medium Voltage Altern Current) connections, section 3.2.

Considering AC connections, the different lay-outs are divided into two families: HVAC (High Voltage Altern Current) and MVAC connections. Within these two AC families, there are also different design options. These options are determined by the number of submarine cables to connect the offshore wind farm with the distribution grid and their voltage level, see section 3.2.1.

With regards to the number of connections to shore, the present evaluation has considered the following configurations: a unique connection of 150 MW, two connections of 75MW (2x75MW) and five connections of 30MW ( 5x30MW).

The voltage level is another important characteristic of the transmission system. Hence, several transmission voltages from medium voltage (36kV) to high voltages (150kV and 220kV) are taken into account in this evaluation. In this way, depending on the N° of clusters and transmission voltage, the suitable cable is used.

In short, the considered AC configurations and their associated cable (

| Cable | Vn (kV) | In(A) | Rac(Ω/Km) | L (mH/Km) | C (µF/Km) |
|-------|---------|-------|-----------|-----------|-----------|
| A | 36 | 911 | 0.0341 | 0.294 | 0.331 |
| B | 150 | 1088 | 0.0205 | 0.352 | 0.233 |
| C | 220 | 1055 | 0.048 | 0.37 | 0.18 |

Table 4.4) are summarized in Table 5.1. Moreover, a couple examples of these configurations are shown in Figure 5.2 and Figure 5.3.

| Cable | Electric configurations |
|-------|------------------------|
| A | 5x30MW-36kV |
| B | 150MW-150kV / 2x750MW-150kV |
| C | 150MW-220kV |

Table 5.1 Cable used for each electric configuration.

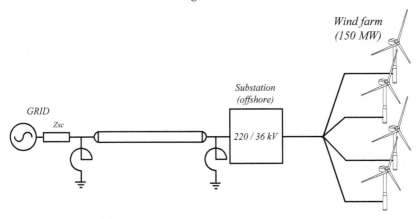

Figure 5.1 Electric configuration of a 150MW-220kV wind farm.

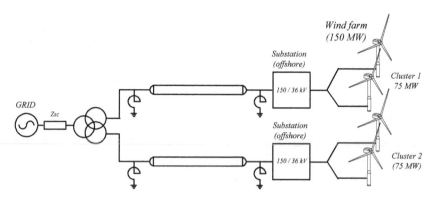

Figure 5.2 Electric configuration of a 2x75MW-150kV wind farm.

### 5.1.1.1 Considered submarine cable model

This evaluation is focused on the selection of the AC transmission lay-out for offshore wind farms. Therefore, the transient behavior of the transmission system has not been taken into account, i.e. the submarine cable has been modeled for steady-state operation.

Thus, the submarine cable has been modeled as a several "π" circuits in series (see section 4.2.2). Due to the fact that this model can represent the cable at steady-state and allows the results measure in intermediate points.

In this chapter, there are taken into consideration three different cables, the same ones of the previous chapter 4, as shown in Table 5.2.

| Cable | Vn (kV) | In(A) | Rac (Ω/Km) | L (mH/Km) | C (μF/Km) |
|-------|---------|-------|------------|-----------|-----------|
| A | 36 | 911 | 0.0341 | 0.294 | 0.331 |
| B | 150 | 1088 | 0.0205 | 0.352 | 0.233 |
| C | 220 | 1055 | 0.048 | 0.37 | 0.18 |

Table 5.2 Submarine cable characteristics.

### 5.1.1.2 Considered reactive power compensation

The reactive power management is an important variable to select a lay-out of the offshore wind farm: This is because with an inadequate compensation the cable can be unusable or causes an increase in conduction losses. Consequently, it is important to determine firstly the reactive power management.

This evaluation does not take into account the technological aspects of how the reactive power is compensated, only the way that it manages the reactive power through the cable. In this way, fixed reactive power compensation at both ends of the cable is considered to perform the management of the reactive power through the submarine cable, see section 4.3.3.

The reason for considering this kind of reactive power management for the submarine cables is because this kind of management reduces considerably the active power losses, at any transmitted power level and for all the layouts, as can be derived from Figure 5.3.

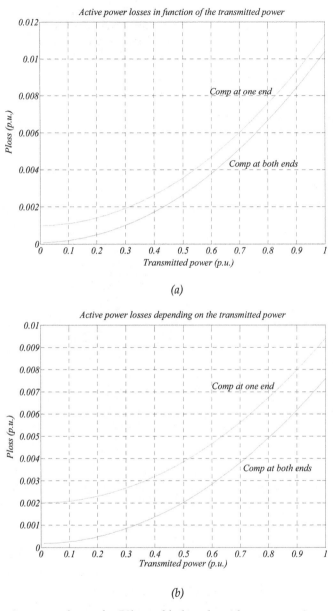

(a)

(b)

Figure 5.3 Active power losses for 50km cable length, with compensation at both ends (red) and onshore compensation (blue). (a) 150MW-150kV and (b) 150MW-220kV.

### 5.1.2 Procedure to calculate the energy transmission cost

As is explained previously, the main criteria to select the electric configuration of the offshore wind farm is economic [69].

Consequently, in this section, for a specific AC offshore wind farm, based on an approach of the cost of the transmission system elements (submarine cable cost, O&M cost...) and depending on these three variables: average wind speed, rated power and location. The most cost effective (economic optimum) lay-out is estimated.

Therefore, the number of cables to use in the transmission system, their voltage level and the characteristics of the offshore platform (if required) are defined. In short, the electrical lay-out.

Notice that the wind turbines are not considered in the present evaluation. The evaluation is focused on the transmission system lay-out, so a simplified model of the entire offshore wind farm is used to estimate the generated energy.

The block diagram of the procedure to calculate the cost of the energy transmission is depicted in Figure 5.4.

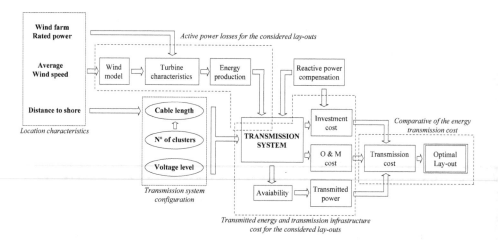

Figure 5.4 Block diagram of the procedure to select the transmission system lay-out.

Firstly, the reactive power compensation of the transmission system is defined (location and quantity). At the same time, the energy production at the offshore wind farm is calculated depending on the rated power of the wind farm and the average wind speed of the location.

In the next step, the active power losses of the transmission system are calculated considering the characteristics of the cable such as: the cable length, the transmission voltage and the transmitted power. The objective to calculate these average power losses is estimate the transmitted energy.

Finally, the transmission cost is defined as the investment cost (estimated operating and maintenance, O&M cost included) of the transmission system divided with the transmitted power. Therefore, energy transmission cost in €/kWh is obtained.

### 5.1.3 Active power losses for the considered lay-outs
Due to the fact that, the energy transmission cost is based on the relation between the investment cost and the generated energy, but measured in the PCC (after the transmission

system), active power losses in the transmission system are a very representative variable to design the transmission layout.

The average active power losses of the transmission system depends on four variables: wind farm's rated power, distance to shore, average wind speed and reactive power management

In the present evaluation, for a 150 MW wind farm with fixed reactive power compensation at both ends, four different configurations are considered.

With regards of wind speed, it can be treated as a continuous random variable. The probability that a wind speed shall occur can be described with a Rayleigh distribution, see section 2.1.2, equation (106).

$$R(v_{wind}) = \frac{k}{c} \cdot \left( \frac{v_s}{c} \right)^{k-1} \cdot e^{-\left( \frac{vs}{c} \right)^k}$$                (106)

Where: $R(v_{wind})$ is probability density, $v_{wind}$ is wind speed (m/s), $k$ is shape parameter (=2) and $c$ is scale parameter.

The average wind speed is chosen from Figure 2.17, an average wind speed for a good location 9 m/s. The Rayleigh distribution for a 9 m/s average wind speed is shown in Figure 5.5.

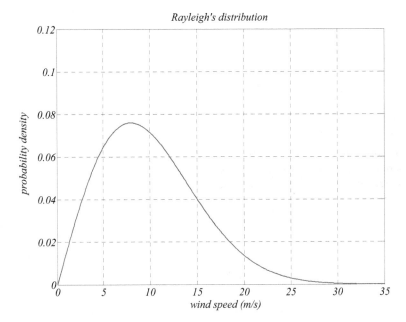

Figure 5.5 Rayleigh distribution for 9 m/s average wind speeds.

On the other hand, in equation (107), the active power produced on the wind turbine (see section 2.2.3) depending on the wind speed and turbine characteristics are calculated (Figure 5.6).

$$P_t(v) = \frac{1}{2} \cdot \rho \cdot \pi \cdot R^2 \cdot V_{wind}^3 \cdot Cp \tag{107}$$

Where: $P_t(v)$=Generated power depending on the wind speed, $\rho$= air density (1.125 measured in kg/m³), $R$=Wind turbine radio (126 m), $V_{wind}$=Wind speed, $Cp$= Aerodynamic efficiency.

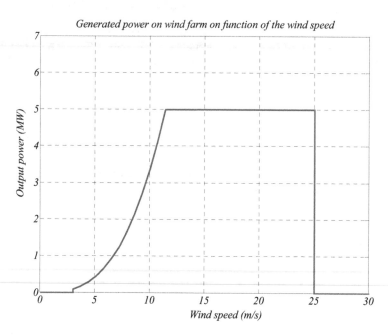

Figure 5.6 Generated power on the wind turbine depending on the wind speed.

Hence, considering the same average wind speed in all the 30 wind turbines (and the same probability density), it is possible to estimate the power generated in the wind farm and its probability density. As a result, it is also possible to obtain the produced average power, equation (108).

$$P_{avg} = \int_{Von}^{Vcut} P(v) \cdot Ry(v) \cdot dv \tag{108}$$

Where: $P_{avg}$=Average input power, $Vcut$= cut out wind speed (30m/s), $Von$= cut in wind speed (3,5m/s), $P(v)$=Input power depending on the wind speed, $Ry(v)$=probability distribution (Rayleigh).

This evaluation is focused in the transmission system, therefore losses in generators and losses in inter-turbine network have not been taken into account. Nevertheless, this simplification does not affects to the analysis, since these losses are the same for all the layouts and only affect to the amount of the produced active power.

Linking the active power losses of the submarine cable depending on the transmitted power level (Figure 5.3), with the produced active power depending on the wind speed (equation (102)). It is possible to obtain the active power losses depending on the wind speed, $P_{loss}(v)$. Finally, using the wind speed distribution probability, the average active power losses can be calculated, equation (109):

$$P_{loss\_avg} = \int_{Von}^{Vcut} P_{loss}(v) \cdot Ry(v) \cdot dv \tag{109}$$

Where: $P_{loss\_avg}$=Average active power losses, $Vcut$= cut out wind speed (30m/s), $Von$= cut in wind speed (3.5m/s), $P(v)$=Input power depending on the wind speed, $Ry(v)$=probability distribution (Rayleigh).

These losses depending on cable length for the considered AC configurations are illustrated in Figure 5.7.

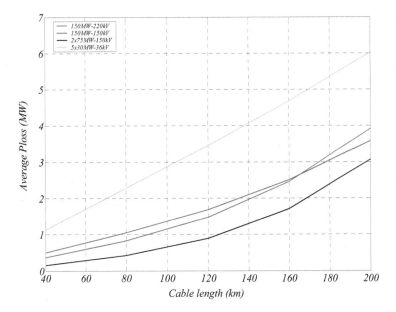

Figure 5.7 Average losses depending on the cable length for different lay-out configurations.

Due to the fact that the reactive power generated by the submarine cable has a quadratic relation with the transmission voltage (see section 4.3), increasing transmission voltage, not always conduction losses are reduced.

Looking at Figure 5.7, for cable lengths longer than 160 Km, the active power losses for 220kV configuration are bigger than the losses for the 150kV configuration.

### 5.1.4 Transmitted energy and transmission infrastructure cost for the considered layouts

Based on the estimation of the produced energy and the average losses of the transmission system, equations (108) and (109), it is possible to calculate the transmitted power through the transmission system for a determined life time.

Therefore, dividing this transmitted power with the sum of the investment cost and operating and maintenance cost, it is possible to obtain the energy transmission cost for the considered layouts.

### 5.1.4.1 Transmitted energy

The submarine cables have a failure rate, i.e. a statistical probability to occur a failure given by the manufacturer. In the present studio, the failure rate is considered 0,1/year/100Km [70]. In the same way, the mean time to repair (MTTR) is considered three months. So, based on this data, it is possible to estimate the availability of the submarine cables, equation (110).

In cases with multiple connections to onshore (wind farm divided into clusters) redundant connections between clusters are not considered. Because, as is reported in [41], the most of existing inter-turbine networks have very little redundancy or none at all. Consequently, the same availability of the cables is considered at the same distance to onshore as a simplification.

$$A_{cable} = \frac{t_{life} - f_{rate} \cdot \dfrac{l}{100} \cdot lt_{years} \cdot t_{repair}}{t_{life}} \tag{110}$$

Where: $A_{cable}$ = Cable availability, $t_{life}$ = Life time (month), $f_{rate}$=Failure rate (failure / 100 km / year), $l$=submarine cable length, $lf_{years}$= Life time (year), $t_{repair}$=MTTR (month).

The considered availability depending on submarine cable length is shown in Table 5.3.

| | 10km | 20km | 40km | 80km | 120km | 160km | 200km |
|---|---|---|---|---|---|---|---|
| Availability (20yrs) | 0.9975 | 0.995 | 0.99 | 0.98 | 0.97 | 0.96 | 0.95 |

Table 5.3 Availability depending on submarine cable length.

For electrical configurations with offshore platform, it is also considered the availability of the step-up transformer. The failure rate is considered 0.03/year with 6 month of MTTR [71]. Therefore, the availability of the transformer can be estimated as follows:

$$A_{trafo} = \frac{t_{life} - f_{rate} \cdot lf_{years} \cdot t_{repair}}{t_{life}} \tag{111}$$

Where: $A_{trafo}$ = Transformer availability, $t_{life}$ =Life time (month), $f_{rate}$=Failure rate (failure/year), $lf_{years}$= Life time (year), $t_{repair}$ =MTTR (month).

Finally, the transmitted energy is calculated, equation (112). The results for the considered layouts calculated with 9 m/s average wind speed and 20 years life time are summarized in Table 5.4.

$$E_{trans} = \left(P_{avg} - P_{loss\_avg}\right) \cdot t_{life} \cdot A \tag{112}$$

Where: $E_{trans}$ =Transmitted energy, $P_{avg}$ =Average generated power, $P_{loss\_avg}$ =Average active power losses, $t_{life}$ = Life time (h), $A$ = Availability.

| Transmitted Energy (MWh) | 10km | 20km | 40km | 80km | 120km | 160km | 200km |
|---|---|---|---|---|---|---|---|
| 150MW - 220kV | 12909668 | 12861624 | 12764211 | 12556214 | 12317201 | 12025728 | 11658289 |
| 150MW - 150kV | 12904426 | 12851165 | 12741316 | 12519127 | 12283552 | 12018159 | 11713215 |
| 2x75MW - 150kV | 12918843 | 12879928 | 12802196 | 12626008 | 12417638 | 12151872 | 11798099 |
| 5x30MW - 36kV | 12875852 | 12794161 | 12631263 | 12307169 | 11982751 | 11652341 | 11306269 |

Table 5.4 Transmitted energy for different configurations and lengths.

### 5.1.4.2 Cost of the connection infrastructure

With regards the cost of the connection infrastructure, in the present evaluation, the cost of the following elements are considered: transformer, offshore platform (if required), submarine cables, installation of submarine cables and reactive power compensators.

- Cost of the offshore platform, estimated at the same way that is done in [72].

$$C_{Plataform} = 2.14 + 0.0747 \cdot P_{rated} \tag{113}$$

$C_{platform}$ = Cost of the offshore platform (M€), $P_{rated}$ = Rated power of the wind farm (MW)

- Cost of the transformer, estimated from equation (114) [68].

$$C_{transformer} = 0.03327 \cdot P_{rated}^{0.7513} \tag{114}$$

$C_{transformer}$ = Cost of the transformer (M€), $P_{rated}$ = Rated power of the transformer (MVA).

- Cost of cable installation, this cost depends on the location of the wind farm and is difficult to estimate accurately. In this evaluation for cases with more than one line to connect to onshore, it is considered that all the submarine cables are installed together. The cost of the installation is estimated at 0.256 M€/Km [72].
- Cost of the submarine cables (estimation). This cost can have high fluctuations depending on the market.

|  | 36kV | 150kV | 220kV |
|---|---|---|---|
| €/km | 200.000 | 500.000 | 600.000 |

Table 5.5 Submarine cable cost.

- Cost of the static reactive power compensation [68].

$$C_{comp} = 0.02218 \cdot Q^{0.7513} \tag{115}$$

$C_{comp}$ = Cost of the reactive power compensation (M€), $Q$ = Reactive power to compensate (MVAR).

### 5.1.4.3 Cost of operating and maintenance

Maintenance of the submarine cables: The reparation costs are increasing and are strongly dependent on the market, due to the fact that a limited number of companies have the capability to install and repair submarine cables. However, in [73], the reparation cost for one XLPE 3-Core 132kV submarine cable is estimated on 3 M£ (3.3 M€) and the reparation cost for one XLPE 3-Core of 220kV to 400kV on 4.4 M£ (4.85 M€).

Thus, in the present section as an approximation, for submarine cables with voltages lower than 132kV (36kV) the reparation cost is considered 3.3 M€ and for submarine cables with voltages higher than 132kV (150kV/220kV) 4.85 M€. Therefore, the results for different cable length considering the failure rate of the submarine cables 0.1 / year / 100km and the life time 20 years is summarized in Table 5.6

| N° of repairs (20yrs) | 10km | 20km | 40km | 80km | 120km | 160km | 200km |
|---|---|---|---|---|---|---|---|
| 150MW - 220kV | 0.2 | 0.4 | 0.8 | 1.6 | 2.4 | 3.2 | 4 |
| 150MW - 150kV | 0.2 | 0.4 | 0.8 | 1.6 | 2.4 | 3.2 | 4 |
| 2x75MW - 150kV | 0.4 | 0.8 | 1.6 | 3.2 | 4.8 | 6.4 | 8 |
| 5x30MW - 36kV | 1 | 2 | 4 | 8 | 12 | 16 | 20 |

Table 5.6 Estimated number of repairs in the life time of the cable (20 years) for different cables and cable lengths.

Maintenance of the offshore transformer: for the step-up transformer on the offshore platform a 0.03 / year failure rate is considered and the life time 20 years. As a result, the failure probability for each platform in its life time is 0.6. The reparation cost for these failures is considered 2.5 M£ (2.75 M€) from [71].

In short, considering the values of the previous sections, the required investment cost for different configurations and lengths are shown in Table 5.7.

| Cost (M€) | 10km | 20km | 40km | 80km | 120km | 160km | 200km |
|---|---|---|---|---|---|---|---|
| 150MW - 220kV | 25.2860 | 34.9340 | 54.1660 | 92.5830 | 130.9920 | 169.3490 | 207.7010 |
| 150MW - 150kV | 24.2360 | 32.8490 | 50.0240 | 84.3430 | 118.6670 | 152.9460 | 187.2240 |
| 2x75MW - 150kV | 29.9970 | 41.9420 | 65.8120 | 113.5160 | 161.1980 | 208.8660 | 256.5260 |
| 5x30MW - 36kV | 15.9400 | 31.8550 | 63.6700 | 127.2700 | 190.8500 | 254.4200 | 317.9800 |

Table 5.7 Required investment cost for different configurations and lengths

### 5.1.5 Comparative of the energy transmission cost in €/kWh for the considered layouts

Considering that the invest cost is paid through the offshore wind farm's life time, the total invest cost (including financial costs) is calculated as follows [68]:

$$C_{invest} = \frac{rate \cdot (1 + rate)^{lf_{years}} \cdot lf_{years}}{\left((1 + rrate)^{lf_{years}} - 1\right)} \cdot Invest \qquad (116)$$

Where: $C_{invest}$ = Total investment cost, $rate$ = Interest rate (4%), $lf_{years}$= Life time (years), $Invest$ = Investment cost today.

Therefore, based on the energy generated and transmitted to the PCC (Table 5.4) and the estimated cost for each one of the transmission configurations (Table 5.7), the transmission cost (€/kWh) is calculated, equation (117). Finally, the results for the considered layouts are shown in Figure 5.8.

$$C_{trans} = \frac{C_{invest}}{E_{trans}}$$ (117)

Where: $C_{trans}$ = Energy transmission cost (€/kWh), $C_{invest}$ = Total investment cost, $E_{trans}$ = Transmitted energy.

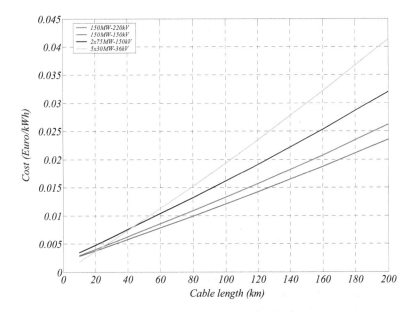

Figure 5.8 Energy transmission cost for different layouts.

In agreement with built wind farms, MVAC transmission systems are the best option near to shore. With short cable lengths (<20Km) MVAC connections are more economic than other layouts, due to the money saved in the offshore platform. On the contrary, with big cable lengths the cables costs do not compensate the money saved in the offshore platform, because submarine cables are very expensive.

The configuration of the two clusters of 75 MW connected with 150kV have the smaller active power losses, but it is not the cheapest option to transmit the energy to shore for any considered length in the considered conditions. The bigger transmitted energy due to the less conduction losses does not compensate the cost of an extra cable.

Considering the cost difference in reactive power compensators and the cable, the option of 150kV option is cheaper than the option of 220kV. What's more, for lengths bigger than 160 Km, the option of 220kV have more conduction losses, increasing the cost difference for long lengths.

Notice that these results are for a location with an average wind speed of 9 m/s and for the considered costs. A strong variation in considered cost estimations can change these results, i.e. these results are an approximation using these cost estimations.

## 5.2 Characterization of a base offshore wind farm

Based on the lay-out selection process of the previous section and the current state of the technology, in the present section the main components of a wind farm are characterized. The objective is the definition of a base scenario to perform based on it several analyses of the critical aspects in the offshore wind farms energy transmission and grid integration.

For that purpose, the physical characteristics of the offshore wind farm, as well as the model used for represent these components are defined. For active elements also their associated control strategy is described.

### 5.2.1 General layout of the considered offshore wind farm

Considering a 150 MW rated power offshore wind farm in a location 50km to the shore (cable length), a HVAC lay-out with a transmission voltage level of 150kV is the most cost effective option, based on the previous section 5.1.

Looking to the current state of wind turbines technology, modern turbines being erected both onshore and offshore are likely to be generating between 1.5 MW onshore and 3MW offshore [74]. More specifically, Vestas has a 3 MW wind turbine, the V90-3MW and Siemens Wind Power has two products, rated at 2.3MW and 3.6MW.

However, the situation is changing. REpower, has recently installed its first offshore turbines, rated at 5MW (soon to be increased to 6MW). Furthermore, Multibrid (Areva) is preparing to install its first turbines offshore, also rated at 5MW [75].

A progression from the G10X 4.5-MW turbine, Gamesa is developing the G11X, a permanent magnet generator with full-scale converter [76]. In the same way, Vestas, is developing the V112-3.0MW Offshore, which also has a permanent magnet generator with a full-scale converter [77]. As the other manufacturers, General electric has the GE 4.0-110 offshore turbine, a permanent magnet generator, with direct-drive technology [78].

In concordance with the plans of the manufacturers, several reports predicted that in the future, the rated power of the offshore wind turbines will be 5MW or bigger [24], [79]. Therefore, following the trend of the wind turbine manufacturers, the considered wind turbines for the base scenario have 5 MW rated power and full-scale converter, see section 2.5.3.4.

With regards to the spatial disposition of the wind turbines, it is considered as a rectangular (see section 3.3). The separation between wind turbines is varied depending on the size of the wind turbines (due to the aerodynamic efficiency) and to avoid overvoltage in each feeder, from 500m to 1000m [40]. In this case, this separation is considered 1000m.

The inter turbine grid, typically medium voltage, has a 33kV voltage (see section 3). The design of the inter-turbine grid is considered radial without any redundant connections. Due to the fact that the most of the offshore wind farms have not these connections [41]. For the considered scenario, each one of those radials is composed by 6 wind turbines of 5MW.

In short, the base scenario used to analyze the critical aspects of the energy transmission is shown in Figure 5.9.

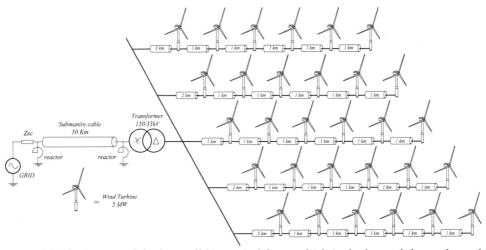

Figure 5.9 The lay-out of the base offshore wind farm, which is the base of the performed analysis.

General description of the main parts of the offshore wind farm:

1. **Wind farm:** Made up with 30 wind turbines of 5 MW.

2. **Inter-turbine grid:** Composed by five feeders of 30 MW (6 wind turbines) connected at 33kV voltage level. The spatial disposition of the grid has a square shape with a separation of 1km between wind turbines. There is not implemented any kind of redundant connections.

3. **The step-up transformer:** The step-up transformer increases the voltage of the inter turbine grid to a suitable voltage to the transmission system. Due to the fact that the inter-turbine grid has a 33kV service voltage and the energy transmission is performed at 150kV.

4. **Collector point:** The point where the energy generated in the wind farm is collected to transmit through the submarine cable to the PCC.

5. **Submarine cable:** The physical medium to transfer the energy from the PC (the collector point) to the PCC (Point of common coupling). This energy is carried at 150kV and through a 50 Km submarine cable.

6. **Reactive power compensation:** Huge reactors placed at both ends of the submarine cable, adjusted to the conditions where the cable is transmitting the rated power. The purpose of those reactors is the management of the reactive power through the cable and the improvement of the energy transmission. Reduces the required rated current of the cable and active power losses.

7. **Point of common coupling (PCC):** The point where the transmission system of the wind farm meets the main grid. The point where the transmission system has to fulfill the grid code requirements (THD, Power factor, LVRT...).

8. **Main grid:** The main distribution grid, in the present case simplified as a ideal voltage source and a sort circuit impedance.

## 5.2.2 Wind turbine

The manufacturer industry is developing the offshore wind turbines technology to bigger rated powers and to permanent magnet generators with direct drive. As a result, the main manufactures have wind turbines in this way: G10X 4.5-MW of Gamesa has a permanent magnet generator and full converter, like GE 4.0-110 of General Electric and The V112 of Vestas.

Thus, as the standard wind turbine for the base scenario is defined a 5MW, full-scale converter wind turbine with permanent magnet generator, Figure 5.10.

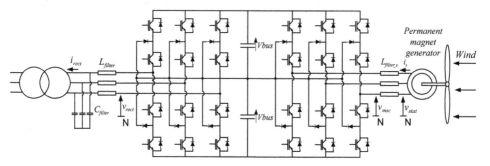

Figure 5.10 Electric scheme of the Neutral-point-clamped full-scale converter wind turbine based on a permanent magnet generator (direct drive).

For the technology of the converter, a neutral-point-clamped topology has been considered. Following the trend of increasing power and voltage levels in wind-power systems [39].

As the power ratings of the wind turbines increases, medium-voltage converters become more competitive. The cost of the cables and connections are reduced at this voltage level and those for the transformer and generator are barely affected. Furthermore, medium voltage converters need fewer components and as a result improve the reliability of them [80].

Therefore, considering this power range (5 MW) and the nature of the application (wind energy) a three-level topology is used with a 3.3kV ($v_{rect}$) output voltage. For the semiconductors, IEGTs (Injection Enhanced Gate Transistor) are used and an 1100 Hz switching frequency.

In short, some features very similar to the Ingedrieve MV 300 of Ingeteam (Figure 5.11), which is used as a reference to characterize the full-scale converter of the considered wind turbine.

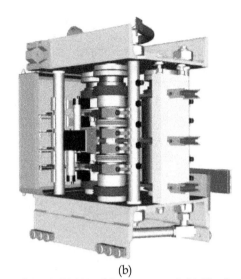

(a) (b)

Figure 5.11 (a) External appearance of the Ingedrive MV 300 of Ingeteam and (b) The basic power module (BPM) equivalent to a branch of the three-level converter, the module is composed by 4 IGBTs, 4 weeling diodes and 2 clamp diodes, all of them based on IEGTs technology.

The general characteristics of the Ingedrive MV 300 of Ingeteam [81] are summarized in Table 7.8.

| Characteristic | Value |
| --- | --- |
| Rated Output Power: | 8 MVA |
| Topology: | NPC 3 Level AFE (press pack IEGTs) |
| Cooling System: | Deionized Water Cooled |
| Line Supply Voltage: | 3300 Vac ±10% |
| Line Supply Frequency: | 50 / 60 Hz ± 5% |
| Rated Output Current: | 1475 Aac |
| Efficiency at 100% of the Rated Operating Point: | 97.8 % |
| PWM Frequency | 1 kHz |

Table 5.8 General characteristics of the ingedrive MV 300 of Ingeteam.

The machine side converter not always can reduce the collected active power or cannot do that reduction as fast as is required. In these cases, the difference between the active power collected by the machine side converter and the evacuated active power by the grid side converter will lead to increase the DC voltage of the BUS.

To solve this problem, i.e. to avoid over voltages in the BUS, the wind turbine is provided with a chopper circuit to consume this exceeded energy.

The DC-link brake chopper allows the wind turbine to keep connected during the grid faults by wasting the active power impossible to evacuate to the grid. As a result, avoids over voltages in the DC-link during the grid faults [82], [83].

In the present case, the control strategy of the DC chopper is On/Off kind, Figure 5.12 (a). Activating and deactivating the DC chopper, in order to maintain the DC-link voltage in a suitable range. In short, if the voltage in the DC-link trespasses the upper limit, the DC chopper is activated to reduce the voltage of the DC-link, on the contrary, when the voltage in the DC-link trespasses the down limit the DC chopper is disconnected.

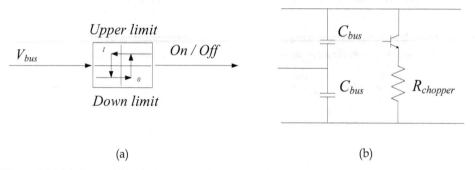

(a)                                                                          (b)

Figure 5.12 (a) Control block diagram (On/Off) of the DC chopper. (b) Electric lay-out of the considered chopper circuit.

### 5.2.2.1 Considered wind turbine model

To model the considered wind turbines some simplifications are performed. Firstly, the grid side and the generator side converters are considered decoupled. Thus, as the analysis is focused on the electrical aspects of the offshore wind farms power transmission, only the grid side converter is considered [39], [84]. The generator-side converter and their respective controllers are not included in the model [85], [88].

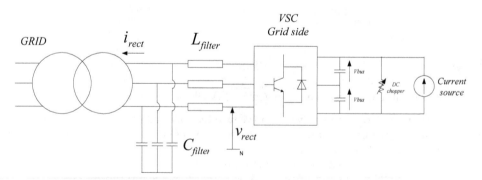

Figure 5.13 Electric scheme of the considered wind turbine model.

### 5.2.2.2 Control strategy of the grid side converter

The wind turbine has its own grid code requirements. These requirements are over the same aspects of the wind farms requirements (THD, LVRT…) and oriented to help the wind farm accomplishing grid codes. One of the most demanding requisite that has to afford a wind turbine is the LVRT requirement [89] (see section 5.2.2.4).

In this way, most voltage dips caused by network faults can be decomposed in positive-negative and zero-sequence components. Thus, it is reasonable to use these symmetrical components in the control of the grid side voltage converter (VSC) [90], [91].

Under a unbalanced voltage dip situation, the active and reactive power exchanged by the converter with the grid, present an oscillatory behavior that can be represented according to the following expressions [92]:

$$P(t) = P_0 + P_{c2} \cdot \cos(2wt) + P_{s2} \cdot \sin(2wt) \tag{118}$$

$$Q(t) = Q_0 + Q_{c2} \cdot \cos(2wt) + Q_{s2} \cdot \sin(2wt) \tag{119}$$

Where, each term of these equations can be represented in the $dq$ frame (see appendix D) as:

$$P_0 = \frac{3}{2}\left(v_d{}^+ i_d{}^+ + v_q{}^+ i_q{}^+ + v_d{}^- i_d{}^- + v_q{}^- i_q{}^-\right) \tag{120}$$

$$P_{c2} = \frac{3}{2}\left(v_d{}^+ i_d{}^- + v_q{}^+ i_q{}^- + v_d{}^- i_d{}^+ + v_q{}^- i_q{}^+\right) \tag{121}$$

$$P_{s2} = \frac{3}{2}\left(v_q{}^- i_d{}^+ + v_d{}^- i_q{}^+ + v_q{}^+ i_d{}^- + v_d{}^+ i_q{}^-\right) \tag{122}$$

$$Q_0 = \frac{3}{2}\left(v_q{}^+ i_d{}^+ + v_d{}^+ i_q{}^+ + v_q{}^- i_d{}^- + v_d{}^- i_q{}^-\right) \tag{123}$$

$$Q_{c2} = \frac{3}{2}\left(v_q{}^+ i_d{}^- + v_d{}^+ i_q{}^- + v_q{}^- i_d{}^+ + v_d{}^- i_q{}^+\right) \tag{124}$$

$$P_{s2} = \frac{3}{2}\left(v_d{}^+ i_d{}^- + v_q{}^+ i_q{}^- + v_d{}^- i_d{}^+ + v_q{}^- i_q{}^+\right) \tag{125}$$

Focusing the previous set of equations to the positive sequence, the $dq$ positive current references can be calculated as:

$$
\begin{bmatrix} i_d{}^{+*} \\ i_q{}^{+*} \\ i_d{}^{-*} \\ i_q{}^{-*} \end{bmatrix}
= \frac{3}{2} \cdot
\begin{bmatrix}
v_d{}^+ & v_q{}^+ & v_d{}^- & v_q{}^- \\
v_d{}^- & v_q{}^- & v_d{}^+ & v_q{}^+ \\
v_q{}^+ & -v_d{}^+ & v_q{}^- & -v_d{}^- \\
v_q{}^- & -v_d{}^- & v_q{}^+ & -v_d{}^+
\end{bmatrix}^{-1}
\cdot
\begin{bmatrix} P_0{}^* \\ 0 \\ Q_0{}^* \\ 0 \end{bmatrix}
\tag{126}
$$

Therefore, based on the reactive power reference ($Q_0{}^*$) and the active power reference ($P_0{}^*$, imposed by the DC bus voltage regulator), the references for the positive sequences of the currents are calculated by using the inverse matrix ($A^{-1}$), equation (126).

As regards to the negative current references, the objective of the converter is to compensate as much as possible the negative sequence components. Consequently, the references for the negative sequences of the currents are always zero.

The current controllers are implemented using two proportional-integral gains in the $dq$ frame with cross-coupled terms ($V_{d\_cct}^{-}$, $V_{d\_cct}^{+}$, $V_{q\_cct}^{-}$ and $V_{q\_cct}^{+}$) for each sequence (positive and negative). To estimate these cross-coupled terms, the LC-L filter is simplified considering it as a L filter [93]. The block diagram of the control strategy is depicted in Figure 5.14.

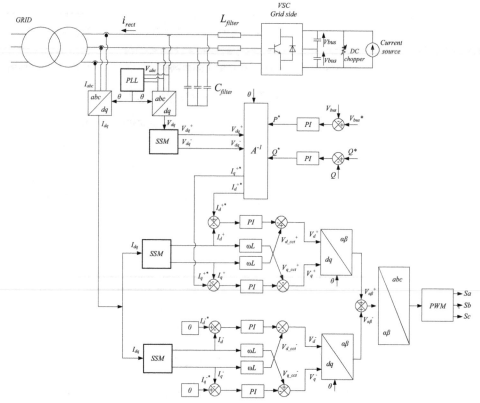

Figure 5.14 The block diagram of the implemented control strategy.

The control strategy is based on positive and negative sequences. Consequently a sequence separation method (SSM) is used to extract the sequences.

As grid side converter, it has to be synchronized with the phase angle of the grid. To achieve this synchronization, a phase locked loop (PLL) working with an SSM is used. In this way, the PLL can guarantee angle precision when asymmetrical grid faults or unbalanced grid conditions occurs [39], [94].

For the implementation of the PLL, the "$d$"-axis of the synchronous reference frame is aligned with the positive sequence vector of the grid voltage.

### 5.2.2.3 Grid side converter's connection filter

To choose an optimal filter topology for the NPC inverter of the offshore wind turbines, the efficiency, weight and volume have to be considered. Due to the fact that in comparison with other application, offshore are difficulties with the transportation and the installation of the filters.

In this way, LCL filters have the advantage of providing a better decoupling between filter and grid impedance (reducing the dependence on grid parameters). In this kind of filters it is also possible the reduction of the cost and weight by increasing the value of the capacitor [95].

Thus, to connect the grid side inverter to the inter-turbine grid a LCL filter is used [95], [96]. The filter of the base offshore wind farm is adjusted with the following criteria:

- The resonant frequency of the filter has to be less than the half of the switching frequency.
- The resonant frequency of the filter has to be at least 10 times bigger than the fundamental frequency.

The characteristics of the considered LC-L filter are summarized in Table 7.9 and the frequency response is shown in Figure 5.15.

| LCL values | Fres |
|---|---|
| 0.8 mH-175uF-0.4mH | 550Hz |

Table 5.9 Characteristics of the LCL filter.

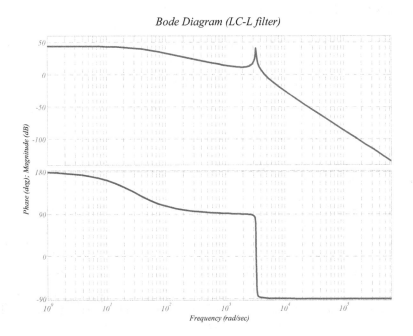

Figure 5.15 Bode diagram of the LC-L filter.

#### 5.2.2.4 Verification that the wind turbine is suitable to connect to the main grid according to the grid code requirements

All the offshore wind farms connected to a transmission grid have to fulfill with the grid code requirements of the system operator (SO). In the present studio, the considered offshore wind farm is connected to a distribution grid operated by REE. So, the offshore wind farm and the wind turbines placed in it have to fulfill the REE grid code requirements (P.O. 12.3).

In this way, for a specific grid code requirements, it is possible to homologate the control strategy and the considered features of the wind turbines. Because, if the proposed wind turbine model fulfills the REE grid code requirements and this fulfillment is verified by using the standard procedure, this model can be considered as a realistic approximation.

Looking into the grid code requirements, the most demanding requisite for wind farms and wind turbines is the Low Voltage Ride Through capability [97], [98] and [99]. Therefore, in the present section this aspect is analyzed to validate the wind turbine model.

#### 5.2.2.4.1 Verification procedure established by REE

For testing and validation wind turbines, i.e. to make sure that the wind turbines fulfill the grid code requirements. REE have defined a procedure detailing all the tests and characteristics in the validation process, the PVVC (Procedimiento de verificación, validación y certificación de los requisitos del PO 12.3 sobre la respuesta de las instalaciones eólicas ante huecos de tensión) [101].

Requirements of the PVVC to validate a wind turbine model

The PVVC specifies in its section 6.2.2 (test conditions for the direct fulfillment of the P.O. 12.3, Particular process) the conditions for each fault category (Table 5.12) for the direct fulfillment of the P.O. 12.3. The PVVC specifies the validation criteria as follows:

- The reactive power consumption of the wind turbine in the zone A of the fault (see Figure 5.16) will not exceed the 15% of its rated power, in 20 ms cycles. On the contrary, in the zone B of the fault, this consumption will not exceed the 5% of its rated power, in 20 ms cycles.

- The net reactive current consumption of the wind turbine after the fault clearance in the zone delimited by $T_3$ and $T_3$+150ms, in 20 ms cycles, must not exceed 1.5 times the rated current, even if the voltage is above the 0.85 pu.

Procedure to define voltage dips or faults:

Based on the IEC 61400-21 standard, the PVVC defines the way to measure the depth of a grid fault, as well as is defined the method to produce the fault. In this way, PVVC specifies that the voltage dip has to be independent to the tested wind turbine. Therefore, the voltage dip is measured in a "no-load" scenario, with the wind turbine disconnected.

According to the PVVC, the voltage dips must be generated using a voltage divider. This divider consists of two inductances in series: The short circuit inductance with a "shock" inductance and a fault or dip inductance, as can be seen in Figure 5.17.

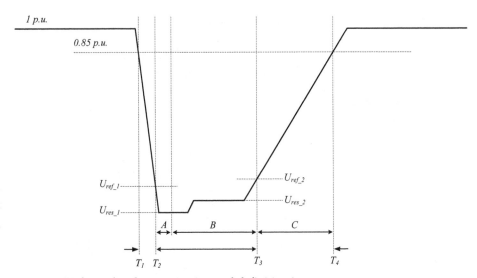

Figure 5.16 Voltage dip characterization and definition in zones.

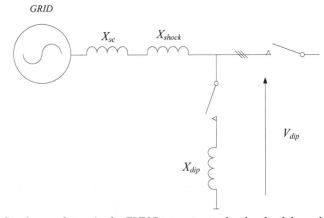

Figure 5.17 Defined test scheme in the PVVC to measure the depth of the voltage fault.

For different dip types, the value of the residuary voltage and the duration are defined in the PVVC as is summarized in Table 5.10

| Dip type | Residuary voltage of the fault ($U_{res}$) | Voltage tolerance (UTOL) | Dip time (ms) | Time tolerance (TTOL, ms) |
|----------|-----------------------------|------------------|-------------|-------------------|
| *Three phase / One phase* | ≤(20%+UTOL) | +3% | ≥(500-TTOL) | 50 |
| *Two phase ungrounded* | ≤(60%+ UTOL) | +10% | ≥(500-TTOL) | 50 |

Table 5.10 Dip voltage characteristics for the test with the wind turbine disconnected.

The value of the residuary voltage is calculated via equation (127):

$$V_{dip}(p.u.) = V_{grid}(p.u.) \cdot \frac{X_{dip}(p.u.)}{X_{dip}(p.u.) + X_{shock}(p.u.) + X_{sc}(p.u.)} \tag{127}$$

Diagram of the simulation scenario

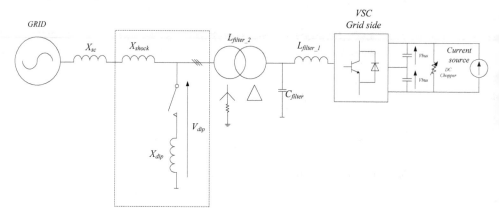

Figure 5.18 Defined test circuit in the PVVC to measure the LVRT capabilities of the wind turbine.

| Parameter | Value |
|-----------|-------|
| $L_{filter\_1}$ | 11,63% |
| $L_{filter\_2}$ | 12% |
| $X_{dip}$ | Depending on the depth of the fault |
| $X_{shock} + X_{sc}$ | 5% |

Table 5.11 Value of the impedances of the simulation scenario.

$X_{shock}$ and $X_{sc}$ impedances have the objective to limit the short circuit current of the grid during faults. As a result, the value of these impedances has specified boundaries in the standard procedure. According to the PVVC, the sum of the both impedances is limited to allow a grids short circuit power equal or bigger than five times the registered rated power of the wind farm (<20%). In the present simulation scenario is chosen 5%.

Considered tests to validate the wind turbine model:

To test and validate the wind turbines, four tests are defined in the PVVC, illustrated in Table 5.12

| Category | Operating Point | Dip type |
|----------|-----------------|----------|
| 1 | Partial Load | 3 phase |
| 2 | Full Load | 3 phase |
| 3 | Partial Load | 2 phase |
| 4 | Full Load | 2 phase |

Table 5.12 Faults and characteristics taken into account in the PVVC.

To be more specific, the PVVC has limited the terms "partial load" and "full load" in a specific operation range, Table 5.13. In the present analysis, there is considered 20% for partial load and 90% for full load.

| | *Registered active power* | *Power factor* |
|---|---|---|
| *Partial Load* | *10%-30% Pn* | *0,90 inductive -0.95 capacitive* |
| *Full Load* | *>80% Pn* | *0,90 inductive -0.95 capacitive* |

Table 5.13 Definition of the operation point ranges before faults.

### 5.2.2.4.2 Results of the considered wind turbine upon the verification procedure

Results of a fault category 1: Three-phase fault - partial load

| | Limit P.O. 12.3 | Test results |
|---|---|---|
| *Net reactive power consumption, in cycles of 20ms, during a period of 150ms after the beginning of the fault:* | -0.1500 | 0 |
| *Net reactive power consumption, during a period of 150ms after the clearance of the fault:* | -0.0900 | 0 |
| *Net reactive current consumption, in cycles of 20ms, during a period of 150ms after the clearance of the fault:* | -1.5000 | -0.2584 |
| *Net active power consumption during the fault:* | -0.1000 | -0.0256 |
| *Net reactive power consumption during the fault:* | -0.0500 | 0 |
| *Fulfillment of the $I_{reactive} / I_{total}$ requirement:* | 0.9000 | 0.9981 |

Table 5.14 Summarized results of a 1st category fault for the test defined in the PVVC.

Results of a fault category 2: Three-phase fault - full load.

| | Limit P.O. 12.3 | Test results |
|---|---|---|
| *Net reactive power consumption, in cycles of 20ms, during a period of 150ms after the beginning of the fault:* | -0.1500 | -0.0467 |
| *Net reactive power consumption, during a period of 150ms after the clearance of the fault:* | -0.0900 | 0 |
| *Net reactive current consumption, in cycles of 20ms, during a period of 150ms after the clearance of the fault:* | -1.5000 | -0.2874 |
| *Net active power consumption during the fault:* | -0.1000 | -0.0266 |
| *Net reactive power consumption during the fault:* | -0.0500 | 0 |
| *Fulfillment of the $I_{reactive} / I_{total}$ requirement:* | 0.9000 | 0.9980 |

Table 5.15 Summarized results of a 2nd category fault for the test defined in the PVVC.

Graphic results:

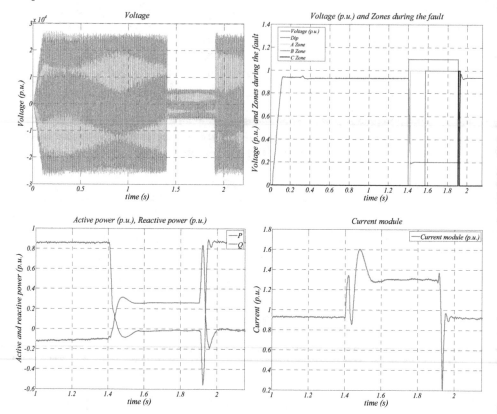

Figure 5.19 Summarized graphical results of a 2nd category fault for the test defined in the PVVC, voltage (module and signals), power and current results.

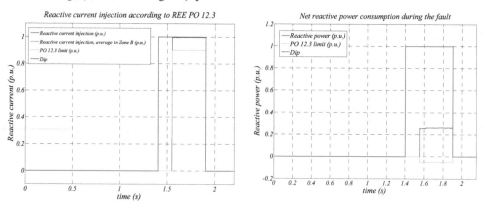

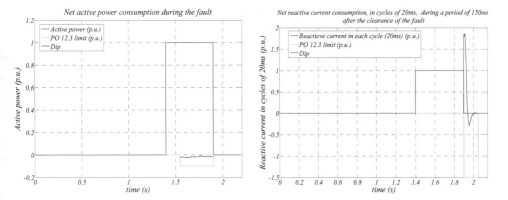

Figure 5.20 Summarized graphical results of a 2nd category fault for the test defined in the PVVC, reactive current and power consumption in B zone and reactive current consumption in C zone.

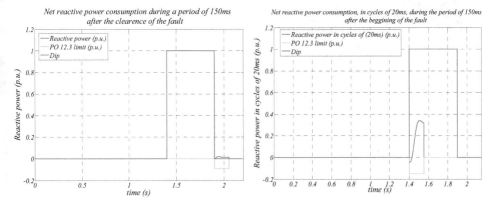

Figure 5.21 Summarized graphical results of a 2nd category fault for the test defined in the PVVC, reactive power consumption in C zone and A zone.

Results of a fault category 3: Two phase ungrounded fault - partial load

|  | Limit P.O. 12.3 | Test results |
|---|---|---|
| Net reactive power consumption, in cycles of 20ms, during the maintenance of the fault: | -0.4000 | 0 |
| Net reactive power consumption, during the maintenance of the fault: | -0.0400 | 0 |
| Net active power consumption, in cycles of 20ms, during the maintenance of the fault: | -0.3000 | 0 |
| Net active power consumption, during the maintenance of the fault: | -0.0450 | 0 |

Table 5.16 Summarized results of a 3rd category fault for the test defined in the PVVC.

Results of a fault category 4: Two phase ungrounded fault - full load

| | Limit P.O. 12.3 | Test results |
|---|---|---|
| *Net reactive power consumption, in cycles of 20ms, during the maintenance of the fault:* | -0.4000 | 0 |
| *Net reactive power consumption, during the maintenance of the fault:* | -0.0400 | 0 |
| *Net active power consumption, in cycles of 20ms, during the maintenance of the fault:* | -0.3000 | 0 |
| *Net active power consumption, during the maintenance of the fault:* | -0.0450 | 0 |

Table 5.17 Summarized results of a 4th category fault for the test defined in the PVVC.

Graphic results:

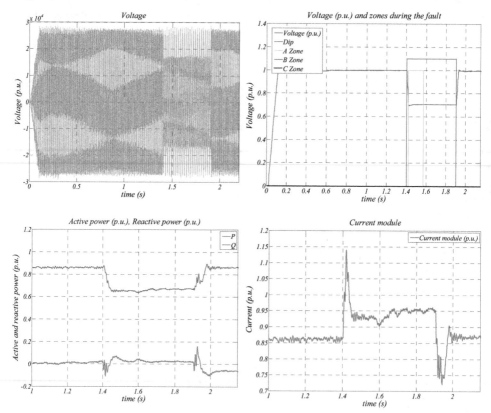

Figure 5.22 Summarized graphical results of a 4th category fault for the test defined in the PVVC, voltage (module and signals), power and current results.

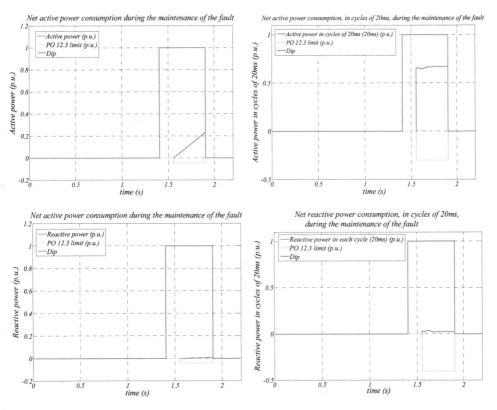

Figure 5.23 Summarized graphical results of a 4th category fault for the test defined in the PVVC, active and reactive power consumption in B zone.

### 5.2.3 Step-up transformer of the wind turbines and the offshore platform

The transformer is modeled as the "classic model" [102], [103]. The electric equivalent diagram is depicted in Figure 5.24

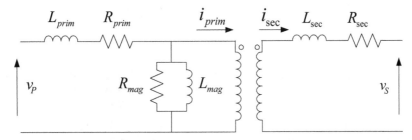

Figure 5.24 Electric equivalent diagram per phase of the transformer model.

The step-up transformer of the base offshore substation has the following characteristics described in Table 5.18

| Parameter | Value |
|---|---|
| Rated power | 150 MVA |
| Primary voltage | 33 kV |
| Secondary voltage | 150 kV |
| Connection | Δ- gY |
| Transformers leakage resistance | 1 % |
| Transformers leakage inductance | 6 % |
| No load losses | 1,78 % |

Table 5.18 Characteristics of the step-up transformer in the offshore substation.

In the same way, the step-up transformer of the wind turbine has the characteristics summarized in Table 5.19

| Parameter | Value |
|---|---|
| Rated power | 5 MVA |
| Primary voltage | 3,3 kV |
| Secondary voltage | 33 kV |
| Connection | Δ- gY |
| Transformers leakage resistance | 1 % |
| Transformers leakage inductance | 6 % |
| No load losses | 1,78 % |

Table 5.19 Characteristics of the step-up transformer in the wind turbine.

### 5.2.4 AC submarine cables

The model and the features of the submarine cables are widely explained in chapter 4. So, in the present section only a resume of the considered submarine cable characteristics is carried out.

The length of the transmission submarine cable for the base scenario is 50 Km and it is modeled using the frequency dependent model in phase domain (section 4.2.2.3.2). The physic characteristics of the transmission cable provided by the manufacturer (Courtesy by General Cable) are shown In Table 5.20.

The other submarine cable used in the base scenario is the medium voltage inter-turbine cable. This cable has to be suitable to connect 6 wind turbines (30MW) of the feeder at 33kV voltage level to the collector point, i.e. suitable to carry at least 525 A.

Therefore, as the inter-turbine submarine cable, an ABB XLPE cable [45] with the adequate nominal voltage and power is chosen. The characteristics of this cable are shown in Table 5.21. This submarine cable is also modeled with the frequency dependent model in phase domain (section 4.2.2.3.2).

| Parameter | Value |
|---|---|
| Rated voltage | 87 / 150kV |
| Rated current | 1088 A |
| Conductors cross section: | 1.200 mm² |
| Separation between conductors: | 97.839996 mm |
| Buried depth | 1 m |
| Shields cross section | 30 mm² |
| Shield type: | Metallic strip |
| Armor type: | Strands crown |
| Diameter of conductor | 43,5 mm |
| Insulation thickness | 20 mm |
| Diameter upon the insulation | 88,5 mm |
| Diameters down the sheath: | 215,6 mm |
| Diameter down the armor: | 226,7 mm |
| Sheath thickness: | 8,9 mm |
| External diameter: | 244,5 mm |
| Relative dielectric constant: | 2,50 |
| Resistivity of the conductor d.c. at 20°C: | 0,0151 Ohm/km |
| Resistivity of the conductor a.c. | 0,0205 Ohm/km |
| Resistivity of the shield d.c. at 20°C: | 0,6264 Ohm/km |
| Nominal capacitance of the cable: | 0,233 µF/km |
| Inductance of the cable: | 0,352 mH/km |

Table 5.20 Cable characteristics provided by General Cable.

| Parameter | Value |
|---|---|
| Nominal voltage | 30kV (36kV) |
| Nominal current | 765 (65°C) – 930 (90°C) A |
| Cross section of conductor | 800 mm² |
| Separation between conductors | 123.65 mm |
| Buried depth | 1 m |
| Shields cross section | 35 mm² |
| Diameter of conductor | 33.7 mm |
| Insulation thickness | 8 mm |
| Diameter upon the insulation | 51.9 mm |
| Relative dielectric constant: | 2,30 |
| Resistivity of the conductor d.c. at 20°C: | 0,02265 Ohm/km |
| Resistivity of the conductor a.c. | 0,024959 Ohm/km |
| Nominal capacitance of the cable | 0,38 µF/km |
| Nominal inductivity of the cable: | 0,31 mH/km |

Table 5.21 Characteristics of the inter-turbine submarine cable [45].

### 5.2.5 The main grid

The main grid is modeled as ideal voltage source and a impedance [39], [83], [84] and [85]. The value of the impedance is calculated considering the point of common coupling (PCC) a strong point [86], [87], with a short circuit power 20 times bigger than the wind farms rated power ($X_{sc}$=5%). Consequently, the short circuit impedance is calculated by equations (128)-(130):

$$S_{sc} = 20 \cdot S_{windfarm} \tag{128}$$

$$S_{sc} = 3 \cdot V_{phase} \cdot I_{sc} = 3 \cdot X_{sc} \cdot I^2_{sc} \tag{129}$$

$$X_{sc} = \frac{S_{sc}}{3 \cdot I^2_{sc}} \tag{130}$$

Where: $S_{sc}$ is the short circuit power of the PCC, $S_{windfarm}$ is the rated power of the wind farm, $I_{sc}$ is the short circuit current and $X_{sc}$ is the short circuit impedance. The short circuit impedance is considered an inductor as a simplification.

### 5.2.6 Protection scheme (Fuses and breakers)

As any other electrical installation, offshore wind farms must be protected against different eventualities in the system. Therefore, in this section, the breakers and fuses considered for the electrical connection infrastructure are defined.

In this way, the considered scenario has fuses in each wind turbine before the step up transformer (*Fuse WT*), see Figure 5.25.

With regards to the breakers, as for the fuses, it is considered one for each wind turbine (*BRK WT*), but in this case, the breakers are placed after the step-up transformer. Furthermore, there are considered auxiliary breakers to allow the disconnection of feeder parts (*BRK aux*) and breakers to disconnect complete feeders (*BRK 3 to BRK 7*).

Finally, in order to provide the offshore wind farm with the capability to disconnect completely from the grid, two general breakers are considered (*BRK 1* and *BRK 2*).

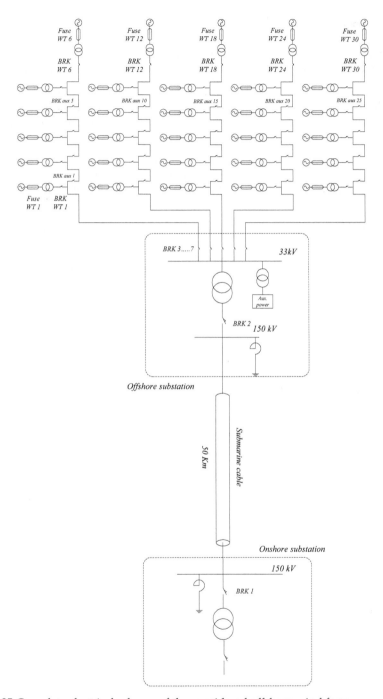

Figure 5.25 Complete electrical scheme of the considered offshore wind farm.

## 5.3 Chapter conclusions

The main objective of this book is the definition of a methodology to design AC offshore wind farms. To focus the problem, in the present chapter, the main elements of the base offshore wind farm are characterized based on the current state of the technology.

The wind turbines are considered a key issue for further analysis. So, after the definition of their rated power, the control strategy and the filter of the grid side converter, the wind turbines are tested via simulation to verify that they can fulfill the grid code requirements and as a result are suitable to place in the offshore wind farm.

Thus, considering this base scenario as a representative case, in the next chapters the key issues of the electric connection infrastructure are evaluated, such as: the frequency response for the electric connection infrastructure or the transient behavior of the offshore wind farm.

# Chapter 6

# Evaluation of Harmonic Risk in Offshore Wind Farms

The interaction between the offshore installations and the onshore grid can cause harmonic amplifications. This aspect is not trivial, because as a result of this harmonic amplification, the harmonic level in the point of common coupling of the wind farm can be unacceptable for the grid code requirements.

Offshore wind farms are connected through a widespread medium voltage submarine cable network and connected to the transmission system by long high voltage cables. Submarine power cables, unlike underground land cables need to be heavily armored chapter 4 and are consequently complicated structures. So, in particular this type of power cables have a relatively larger shunt capacitance compared to overhead lines which make them able to participate more in resonant scenarios [25].

The present chapter evaluates the frequency behavior of the offshore wind farms at normal operation (steady state), depending on the design procedure parameters like: the cable length / characteristics, transformers connection and leakage inductance or inter-turbine grids configuration. The analysis is performed from the point of view of the wind turbines, considering them as potential harmonic sources. Thus, the knowledge of the frequency behavior of the offshore wind farm can help to avoid as much a possible the harmonic amplification, at the design stage of the wind farm. This presents new challenges in relation to understanding the nature, propagation and effects of the harmonics.

## 6.1 Harmonics in distribution grids

Nowadays, the state of the distribution grids is significantly different in comparison with the state of two decades ago. The main reason is the existence of no lineal loads. These no lineal loads provoke disturbances, like a high level harmonics in current and voltages [104].

In the same way, there is consolidating a distributed generation system for the distribution grids. This kind of grids have a combination of many types of generation plants, such as cogeneration, combined cycle, wind farms, photovoltaic…Thus, if the distribution grid is made up with many little and medium generation plants, the waveform of the voltage may be distorted.

In conclusion, the electric transmission system is evolving to a scenario with multiple harmonic sources. So, the frequency analysis of the electric grids is becoming an important tool, because can help to improve their efficiency reducing the power associated to these disturbances.

The current and voltage harmonics superimposed to the fundamental wave causes several negative effects in the devices connected to the distribution grid.

The harmonic currents are the cause of the distortion of the voltage wave in different points of the distribution system, i.e. the circulation through the electric grids of these currents provokes distorted voltage drops, so, at the system nodes there are not pure sine waves. Thus, the bigger are the harmonic currents of the power system, more distorted are the voltages in the nodes and bigger the negative effects caused by them.

Therefore, the system operators of the energy distribution grids have specific rules to limit the harmonic emission, for both of them, voltage and current.

## 6.2 Main disturbances caused by current and voltage harmonics

The distorted voltages are the cause of many negative effects to the devices connected to the system. These effects are: The reduction of the devices lifetime, the degradation of the efficiency and the degradation of the operation in general.

The negative effects caused by the harmonics depends on the type of the load and these negative effects can be divided into two groups [105]:

o   Instantaneous effects.

o   Long-term effects due to the heating.

**Instantaneous effects:**

- Displacement of the zero crossing of the voltage wave and because of this the switching conditions of the thyristors.

- Additional errors in induction disks of the electric meters.

- Vibrations and noise, especially in electromagnetic devices (transformers, reactors, etc...).

- Pulsating mechanical torques, due to changes in the value of the instantaneous current.

- Malfunction of the protection devices, like fuses, breakers and digital equipment's for protection [106]. In the case of a system protected against the overvoltage, where the protections are designed to operate with sinusoidal voltages, cannot operate correctly with non-sinusoidal waveforms. The operation can go wrong, from the overprotection of the system to the non-protection of it.

**Long-term effects:** The main long-term effect of the harmonics is the heating of the devices.

Heating of the capacitors: The losses caused by the harmonics are transformed into heat. In the specific case of the capacitor, these losses are: conduction losses and hysteresis losses in the dielectric.

Heating due to additional losses in machines: There are additional losses at the stator (cupper and iron) and at the rotor (magnetic circuit and the coil). These losses are caused by the speed difference in the inductive rotating field between the rotor and stator.

Heating of the transformers: Additional losses due to the skin effect, there is a increment of the resistance for harmonic currents. In the same way, there are also additional losses in the magnetic circuit (Foucault currents) and hysteresis.

<u>Heating of the cables:</u> In cables where are circulating harmonic currents, there are additional losses due to:

- An increment in the apparent resistance due to the skin effect.
- An increment in the effective current for the same active power.
- An increment in the dielectric losses in the insulation with the frequency, if the applied voltage to the cable is significantly distorted.

In Table 6.1, extracted from [107], the negative effects caused by harmonics to electric devices are summarized.

| Effects of the harmonics | Cause | Consequence |
|---|---|---|
| Upon the conductors | • An increment of the Irms.<br>• The skin effect, which reduces the effective area of the conductor. | • Malfunction of the protections.<br>• Overheating of the conductors. |
| Upon the transformers | • An increment of the Irms.<br>• Increments in the Foucault losses due to there are proportional to the square of the frequency. | • An increment of the heating of the coils due to the Joule effect.<br>• An increment of the losses in the iron. |
| Upon the capacitors | • A decrement of the capacitors impedance with the frequency. | • Premature aging and the amplification of the harmonics. |

Table 6.1 Negative effects caused by harmonics to electric devices.

In order to prevent these disturbances, the elimination of all the harmonic components to obtain a pure sine wave is impossible. However, it is possible to achieve a good approximation to a sine wave. For that purpose, there are operation standards and grid codes.

One of those is the IEEE-519 standard, in this standard there are set up the limits of the harmonic amplitudes for current and voltage. For the specific case of the main distribution grid, the harmonic limits are shown in Table 6.2 and Table 6.3.

IEEE-519 Harmonic current limits for HV systems.

| $I_{SC}/I_L$ | $n < 11$ | $11 \leq n \leq 17$ | $17 \leq n \leq 23$ | $23 \leq n \leq 35$ | $35 \leq n$ | THD |
|---|---|---|---|---|---|---|
| <20 | 2.0 % | 1.0 % | 0.75 % | 0.3 % | 0.15 % | 2.5 % |
| 20-50 | 3.5 % | 1.75 % | 1.25 % | 0.5 % | 0.25 % | 4.0 % |
| 50-100 | 5.0 % | 2.25 % | 2.0 % | 0.75 % | 0.35 % | 6.0 % |
| 100-1000 | 6.0 % | 2.75 % | 2.5 % | 1.0 % | 0.5 % | 7.5 % |
| >1000 | 7.5 % | 3.5 % | 3.0 % | 1.25 % | 0.7 % | 10.0 % |

IEEE-Maximum odd harmonic currents for main distribution system, for 69,001kV to 161kV.

Table 6.2 Harmonic current limits for high voltage systems.

IEEE-519 Voltage limits.

| IEEE-Voltage distortion limits. | | |
|---|---|---|
| BUS voltage at the PCC | Individual harmonics | THD |
| 69kV and smaller | 3.0 % | 5.0 % |
| 69,001kV to 161kV | 1.5 % | 2.5 % |
| Over 161kV | 1.0 % | 1.5 % |

Table 6.3 Harmonic voltage limits for high voltage systems.

## 6.3 Frequency response of the transmission system via PSCAD simulation

### 6.3.1 Frequency response of the submarine cable

The frequency response of the submarine cable described in this chapter is based on the model of the submarine cable analyzed and validated in chapter 4. Thus, the present analysis considers this model as a reasonably accurate approximation.

One option to carry out the frequency response analysis in an easy way is the use of the impedance meter provided by PSCAD in its standard library, but this impedance meter have not taken into account completely the validated model.

As is highlighted in the PSCAD user's manual [64], this impedance meter cannot correct the curve fitting errors during the simulation. Thus, in the present analysis, the simulation scenario depicted on Figure 6.1 is used. The simulation of this scenario takes into account the complete cable model and in consequence the results are intended to be more accurate.

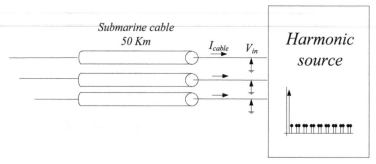

Figure 6.1 Simulation scenario to obtain the frequency response of the submarine cable.

As regards to this scenario, to calculate the impedance of the submarine cable depending on the frequency, a harmonic voltage source is used. So, the harmonic voltage source applies a harmonic train to the submarine cable, which is connected as a load. The cable is the same of the used in chapter 4, its characteristics are shown in Table 4.2.

The harmonic train of input voltage ($V_{in}$), is composed by sinusoidal components in the range of frequencies: 50-5000Hz. The amplitude of these harmonic voltages is 10% of the fundamental (50Hz-150kV). Starting from the 50Hz, the harmonic train has voltage components separated 10Hz one from other, as illustrated in Figure 6.2. These input harmonics in a simplified way can represent the effect of the harmonics generated by the wind turbines, when they are generating energy from the wind.

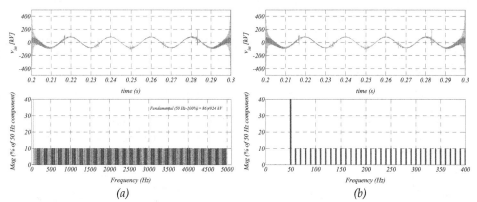

Figure 6.2 Harmonic voltage train applied to the submarine cable model. Resolution 10 Hz.

Measuring the current at the PCC ($I_{pcc}$) and performing the FFT (Fast Fourier Transform) of the signal, Figure 6.3, it is possible to obtain the impedance of the transmission system for each one of the excited frequencies, i.e. it is possible to obtain the evolution of the impedance depending on the frequency.

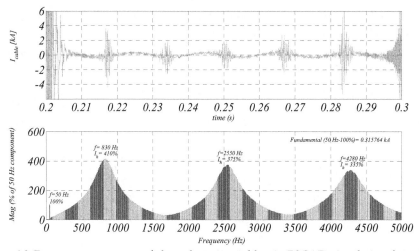

Figure 6.3 Frequency response of the submarine cable via PSCAD simulation for a 50 Km cable, resolution 10 Hz.

Looking at the results depicted in Figure 6.3, the specified cable has several frequencies where the harmonics are amplified. In the analyzed range (50-5000Hz), there are three harmonic groups: around 830Hz, around 2550 Hz and around 4280Hz.

### 6.3.2 Frequency response of the transmission system via PSCAD simulation

The transmission system is the part of the offshore wind farm which makes possible the energy transmission from the collector point (offshore) to the point of common coupling (onshore), in other words, the physic medium to transfer the energy from the wind farm to the main grid and all the support devices.

The transmission system is made up by the step-up transformer, the submarine cable, reactive power compensation elements (if required), and the support devices to integrate the energy in the main grid (if required).

The knowledge of the frequency response of the transmission system and the influence of each component upon this frequency response can help to avoid undesired resonances and harmonics. For that purpose, firstly, in this section the simplest lay-out for the transmission system (transformer, cable and grid, Figure 6.4) is considered, i.e. the necessary elements to perform the energy transmission, without the support devices to improve the transmission.

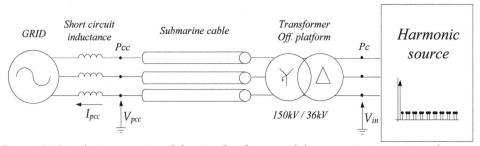

Figure 6.4 Simulation scenario of the simplest lay-out of the transmission system: the step-up transformer, the submarine cables and the distribution grid.

In this case also is performed the same procedure to obtain the frequency response used in the previous section (6.3.1).

To model the grid in a simple manner, a voltage source and short circuit impedance is used. Its characteristics are summarized in Table 6.4. The transformer's connection is $\Delta$- gY, while its characteristics are shown in Table 6.5. Finally, the cable characteristics and cable model are the same of the section 6.3.1

| Parameter | Value |
|---|---|
| Nominal power (Pn) | 150 MW |
| Nominal voltage (Vn) | 150 kV |
| Short circuit inductance | 5 % |

Table 6.4 Characteristics of the main grid.

| Parameter | Value |
|---|---|
| Rated power | 150 MVA |
| Primary voltage | 33 kV |
| Secondary voltage | 150 kV |
| Connection | $\Delta$- gY |
| Transformers leakage resistance | 1 % |
| Transformers leakage inductance | 6 % |
| No load losses | 1,78 % |

Table 6.5 Characteristics of the step-up transformer.

The frequency response of the described transmission system layout is depicted in Figure 6.5.

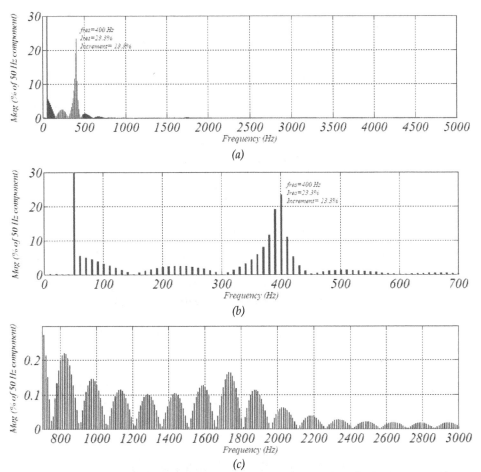

Figure 6.5 Frequency response of the transmission system with only: step-up transformer and submarine cables. FFT of the current at PCC. (a) Whole analyzed spectrum, (b) more detail in the main resonance and (c) more detail in high frequencies.

Looking at Figure 6.5, it is possible to observe that all the multiples of the 3rd order harmonics generated in the wind turbines, cannot trespass to the PCC. This occurs because between these points is placed a transformer with star (grounded)-delta connection.

The transmission system is composed with several inductive components, like the transformer or the short circuit impedance of the main grid. This inductive impedances provokes a significant attenuation of the high frequencies, as can be seen in Figure 6.5 (c), thus, the high frequency harmonic voltages do not affect to the current of the PCC. In fact, in the present analysis, the harmonics higher than 700Hz almost do not affect to the current at PCC.

However, the interaction of the inductive component of the transmission system with the capacitive component of the submarine cable provokes a resonance at 400Hz, becoming these frequencies which are around the 400Hz potentially problematic.

### 6.3.3 The effect of the main components in the frequency response of the transmission system

The analysis of how affects each one of the elements of the transmission system in its frequency response is the first step to avoid undesired resonances and optimize the transmission system design.

Therefore, this section analyses the frequency response of the transmission system varying the characteristics (impedance) of its three main components:

        o   The leakage impedance of the step-up transformer.
        o   The impedance of the submarine transmission line (variation of the cable length).
        o   The short circuit impedance of the main grid.

Firstly, the influence of the step-up transformer is evaluated. Based on the scenario illustrated in Figure 6.4 and applying the same harmonic train (Figure 6.2), the frequency responses of the transmission system are obtained. In this first case, the transformer's leakage inductance has a variation from 3% to 12%, the results are depicted in Figure 6.6.

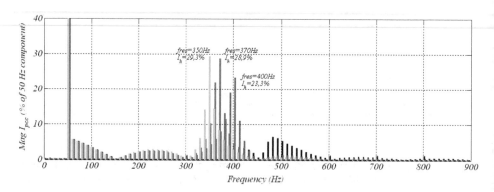

Figure 6.6 Frequency response of the transmission system varying the leakage inductance of the step-up transformer from: 3% (black), 6% (blue), 9% (red) and 12% (green).

As is shown in Figure 6.6, as the leakage inductance of the step-up transformer increases, the frequency of the resonance decreases (from 450Hz to 350Hz).

For the specific case where the leakage inductance is 3%, it is possible to see how the transformer connection does not allows to cross to the PCC the harmonics close to the resonance, Figure 6.7. The resonance is still there (450Hz), but, there are not harmonics to be amplified.

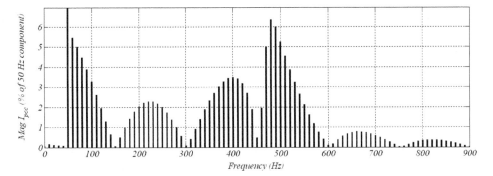

Figure 6.7 Frequency response of the transmission system with a leakage inductance of 3% of the step-up transformer.

The harmonic train used for this analysis has components into de 50-5000Hz range, but not continuously in all this range, the harmonic source generates harmonic voltages in steps of 10 Hz. Thus, using the harmonic train is possible to determinate the resonance with 10 Hz accuracy, i.e. the system has a 10 Hz accuracy

With regards to the amplitude of the resonance, this varies very quickly in few Hz close to the resonance frequency. Consequently, if the harmonic resonance matches up with the exact resonance frequency, the measured amplitude in the simulation will be bigger than in cases where the harmonics in the train are close to the exact frequency of the resonance. Thus, this analysis can measure accurately the frequency of the resonance, but not the amplitude, the amplitude is only an approximated value.

In the next step of the analysis, the influence of the cable length in the range of 20Km to 110Km is evaluated. The frequency response of the considered transmission system with this variation is shown in Figure 6.8.

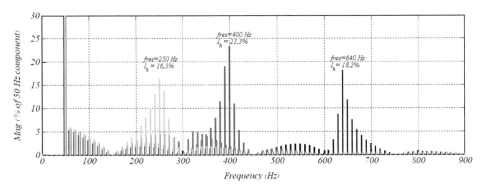

Figure 6.8 Frequency response of the transmission system varying the cable length from: 20Km (black), 50Km (blue), 80Km (green) and 110Km (red).

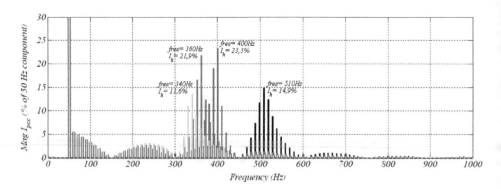

Figure 6.9 Frequency response of the transmission system varying the short circuit impedance from: 2 % (black), 5 % (blue), 8 % (green) and 11 % (red).

In this case, as the submarine cable length increases, the resonance frequency decreases. Note that the resonance of the transmission system with 80Km cable disappears, because in this case also all the multiples of the 3rd order harmonics cannot trespass the transformer.

In the third and last case there are considered different values for short circuit impedances. This variation is from the 2 % to 11 %, the simulation results are depicted in Figure 6.9.

In this last case, increasing the short circuit impedance decreases the resonance frequency, i.e. as in the two previous cases, increasing the inductive impedance or the capacitive impedance the frequency of the resonance decreases.

In the analyzed cases, the biggest variation is between 640Hz-250Hz, caused varying the cable length from 20Km to 110Km. However, in most of the cases the resonance is between 450Hz and 250Hz. In concordance with these results, in [109] is highlighted that AC transmission systems in conjunction with step-up transformer of the offshore substation, present the risk to amplify harmonics at low frequencies (inherently 3rd, 5th and 7th order harmonics).

### 6.4 Frequency response of the transmission system via analytic calculus

The objective of this section is to estimate the main resonance frequency in a simple and accurate way, alternatively to the method described in the previous section. Thus this section studies the calculation of the first resonance frequency of the transmission system, which is the main characteristic of the frequency response, using two different analytic ways.

To characterize in an easy way the main resonance frequency, with a potential risk of harmonic amplification, in [65] is presented a simple method. This approximation only takes into account the capacitive component of the submarine cable, neglecting the resistive and inductive components. In this way, it is possible to simplify the whole transmission system as an equivalent RLC circuit. Then, the resonance frequency of this simplified RLC circuit serves to approximate the resonance of the transmission system.

The second method uses state-space equations to estimate the resonance frequency of the transmission system. These equations take into account all the components of the cable and the short circuit impedance of the main grid, with the advantage that is not too more complicated than the first method.

Finally, to validate these two methods, the results obtained via analytic calculus are compared with the results obtained in simulation with PSCAD as described in section 6.3.2.

### 6.4.1 Frequency response of the transmission system via state space equations

### 6.4.1.1 State space equations for the transmission system with a cable modeled with a unique "π" circuit

At first, in order to explain with an example the method of the state-space equations, the simplest case is analyzed. The step-up transformer is considered as an equivalent inductance and the main grid as an ideal voltage source with short circuit impedance.

With regards to the submarine cable, this is modeled using several "π" circuits in series, (see section 4.2.2.2.4). This model has a frequency limit to represent the cable, i.e. the cable model has a valid range in frequency, out of this frequency range, the cable model and as a result the state-space equations cannot be used, since the error becomes too high. For the simplest case, the present case, the cable is modeled as a unique "π" circuit (N=1).

Once the equivalent circuit in impedances of the considered model is determined, it is possible to obtain the frequency response applying the state-space equations, the procedure is as follows:

In the first step, the names and the directions for all the currents of all the branches of the circuit are established as illustrated in, Figure 6.10.

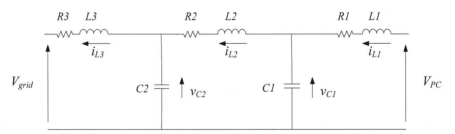

Figure 6.10 Single phase representation of the transmission system with the submarine cable modeled as a unique "π" circuits.

Where: $L1$ represents the equivalent inductance of the step-up transformer, $R1$ represents the equivalent resistance of the step-up transformer, $R2$ represents the resistive part of the submarine cable, $L2$ represents the inductive part of the submarine cable, ($C1=C2$) represent the capacitive part of the submarine cable and ($L3$ and $R3$) represent $Lsc$ and $Rsc$ respectively, short circuit impedances.

In the second step, the differential equations for the currents in inductances and for voltages in capacitors are obtained, equations (131) - (135).

$$\frac{di_{L1}}{dt} = \frac{1}{L1} \cdot \left(V_{PC} - v_{C1} - i_{L1} \cdot R1\right) \tag{131}$$

$$\frac{dv_{C1}}{dt} = \frac{1}{C1} \cdot \left(i_{L1} - i_{L2}\right) \tag{132}$$

$$\frac{di_{L2}}{dt} = \frac{1}{L2} \cdot \left(v_{C1} - v_{C2} - i_{L2} \cdot R2\right) \tag{133}$$

$$\frac{dv_{C2}}{dt} = \frac{1}{C2} \cdot \left(i_{L2} - i_{L3}\right) \tag{134}$$

$$\frac{di_{L3}}{dt} = \frac{1}{L3} \cdot \left(v_{C2} - V_{grid} - i_{L3} \cdot R3\right) \tag{135}$$

Looking at equations (131) - (135), the variables of the differential equations $i_L$ and $v_C$ are independent. In the same way, these variables represent independent physical elements, so, those variables are state space variables.

Thus, if these equations are written in matrix notation (equation (137)), the state-space matrix is obtained as follows:

$$\overset{\bullet}{x} = Ax + Bu \tag{136}$$

$$d/dt \cdot \begin{bmatrix} i_{L1} \\ v_{C1} \\ i_{L2} \\ v_{C2} \\ i_{L3} \end{bmatrix} = \begin{bmatrix} -R1/\!L1 & -1/\!L1 & 0 & 0 & 0 \\ 1/\!C1 & 0 & -1/\!C1 & 0 & 0 \\ 0 & 1/\!L2 & -R2/\!L2 & -1/\!L2 & 0 \\ 0 & 0 & 1/\!C2 & 0 & -1/\!C2 \\ 0 & 0 & 0 & 1/\!L3 & -R3/\!L3 \end{bmatrix} \cdot \begin{bmatrix} i_{L1} \\ v_{C1} \\ i_{L2} \\ v_{C2} \\ i_{L3} \end{bmatrix} + \begin{bmatrix} 1/\!L1 & 0 \\ 0 & 0 \\ 0 & 0 \\ 0 & 0 \\ 0 & -1/\!L3 \end{bmatrix} \cdot \begin{bmatrix} V_{PC} \\ V_{grid} \end{bmatrix} \tag{137}$$

Finally, the poles or eigenvals of the system are calculated (from $A$ matrix), to determine its resonance frequency.

### 6.4.1.2 State space equations for the transmission system with a cable modeled with N "π" circuits

In the next step forward of the analysis, the procedure explained in the previous section (6.4.1.1) is applied to a generic case where the transmission system has a cable modeled with N "π" circuits, Figure 6.11.

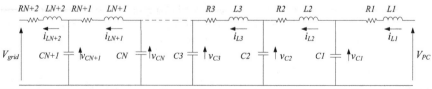

Figure 6.11 Single phase representation of the transmission system with the submarine cable modeled as N "π" circuits.

Where: $L1$ represents the equivalent inductance of the step-up transformer, $(R2=R3=\ldots =RN+1)$ represent the resistive part of the submarine cable, $(L2=L3=\ldots =LN+1)$ represent the inductive part of the submarine cable, $(C1$ to $CN+1)$ represent the capacitive part of the submarine cable and $(LN+2$ and $RN+2)$ represent $Lsc$ and $Rsc$ respectively, short circuit impedances.

For the generic transmission system, following the procedure explained in the previous section, the state-space variables are defined and the estate-space equations are obtained. These state-space equations in matrix notation are displayed in equation (138). The reader can find the similarities of the matrix structure in expressions (137) and (138).

$$
d/dt \cdot
\begin{bmatrix}
i_{L1} \\
v_{C1} \\
i_{L2} \\
\ldots \\
i_{LN+1} \\
v_{CN+1} \\
i_{LN+2}
\end{bmatrix}
=
\begin{bmatrix}
-R1/L1 & -1/L1 & 0 & \ldots & 0 & 0 & 0 \\
1/C1 & 0 & -1/C1 & \ldots & 0 & 0 & 0 \\
0 & 1/L2 & -R2/L2 & \ldots & 0 & 0 & 0 \\
\ldots & \ldots & \ldots & \ldots & \ldots & \ldots & \ldots \\
0 & 0 & 0 & \ldots & -RN+1/LN+1 & -1/LN+1 & 0 \\
0 & 0 & 0 & \ldots & 1/CN+1 & 0 & -1/CN+1 \\
0 & 0 & 0 & \ldots & 0 & 1/LN+2 & -RN+2/LN+2
\end{bmatrix}
\cdot
\begin{bmatrix}
i_{L1} \\
v_{C1} \\
i_{L2} \\
\ldots \\
i_{LN+1} \\
v_{CN+1} \\
i_{LN+2}
\end{bmatrix}
+
\begin{bmatrix}
1/L1 & 0 \\
0 & 0 \\
0 & 0 \\
\ldots & \ldots \\
0 & 0 \\
0 & 0 \\
0 & -1/LN+2
\end{bmatrix}
\cdot
\begin{bmatrix}
V_{PC} \\
V_{grid}
\end{bmatrix}
\quad (138)
$$

### 6.4.1.3 Frequency response of the transmission system via state space equations

Finally, the frequency response of the transmission system with the submarine cable, the step-up transformer and the main grid of the previous section (Table 4.3, Table 6.4 and Table 6.5) via state-space equations is obtained.

For the submarine cable model, 10 "π" circuits in series are considered. In this way, the cable model is composed by sufficient "π" circuits to make possible the representation of the submarine cable in the correct frequency range, i.e. sufficient to represent correctly the cable until the resonance. In more detail, with 10 "π" circuits it is possible to represent the submarine cable in a valid range for all the resonances analyzed in the previous section [54] ,by means of equation (139).

$$
f_{max} = \frac{Nv}{8l} = \frac{N}{8 \cdot l \cdot \sqrt{LC}} = \frac{10}{8 \cdot 110 \cdot \sqrt{0.352 \cdot 0.233 \cdot 10^{-9}}} = 1250\,Hz \Rightarrow 7884\,rad/s \quad (139)
$$

If the resonance frequency estimated in this way is out of the cable model valid range, it is not valid and the analysis must be repeated with a valid cable model.

Finally, applying the developed generic equation (138), to the considered transmission system, the frequency response depicted in Figure 6.12 is obtained.

As can be observed in Figure 6.12, the first resonance of the system is located at 388 Hz, very close to the 400 Hz estimated via PSCAD simulation. With regards to the amplitude of the resonance, in this case also is an approximation, due to the fact that the calculus is based on a model with lumped parameters, not in a model with distributed parameters which is more accurate.

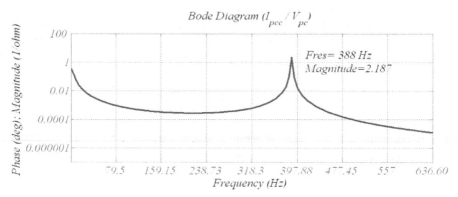

Figure 6.12 Frequency response of the transmission system with the cable modeled as 10 "π" circuits in series using state space equations.

### 6.4.2 Simplified method avoiding the inductive part of the submarine cable

The other method used to compare with the state-space equations is the simplified RLC method, described in [109]. Thus, a comparative is performed comparing the resonance frequency estimated with these three methods for different transmission system. Similarly as done in section 6.3.3, but in this case, the analysis only varies the cable length and the equivalent inductance of the step-up transformer.

The short circuit impedance of the main grid is not taken into account because the simplified RLC method does not consider it. The resonance frequency for the simplified RLC method of [109] is estimated with the following equation (140).

$$f_{Resonance} = \frac{1}{2 \cdot \pi \cdot \sqrt{\left(L_{Transformer} + L_{Cable}\right) \cdot C_{Cable}}} \tag{140}$$

### 6.4.3 Comparison and validation of the frequency response via state space equations

The objective of this section is to validate the state-space equations based method to estimate the first resonance. For that purpose, a comparative of three different methods is carried out. The method based on PSCAD simulation with the validated cable is considered as the most accurate method.

Hence Table 6.6, summarizes the obtained resonance frequencies varying the equivalent inductance of the step-up transformer for the considered three methods. The variation of the equivalent inductance is the same of the section 6.3.3. (3% to 12%).

| Transformers L | 3 % | 6 % | 9 % | 12 % |
|---|---|---|---|---|
| PSCAD simulation | 450 | 400 | 370 | 350 |
| State space equations | 453,6 | 388,3 | 356,5 | 339 |
| Simplified RLC | 264,75 | 218 | 190,9 | 171,4 |

Table 6.6 Comparative of the results obtained for the variation of the equivalent inductance of the step-up transformer.

To verify the state-space equations method at different conditions, a second comparative is carried out. In this second case, the cable length is varied, yielding the resonances depicted in Table 6.7

| Cable length (Km) | 20 | 50 | 80 | 110 |
|---|---|---|---|---|
| PSCAD simulation | 640 | 400 | 300 | 250 |
| State space equations | 631,8 | 388,3 | 297,6 | 246,6 |
| Simplified RLC | 392 | 218 | 156,67 | 124,8 |

Table 6.7 Comparative of the results obtained for the variation of the cable length.

Looking at Table 6.6 and Table 6.7, it can be concluded that the state-space equation method is a good approximation for estimating the resonance frequency, even under different transmission conditions options.

On the other hand, the simplified RLC method does not provide as accurate results as expected to characterize the resonance frequency. However it could be useful to obtain a very simplified and first approximated value.

## 6.5 Frequency response of the offshore wind farm

As is distinguished in chapter 3, the electric infrastructure of the offshore wind farm's connection is divided into two parts: The transmission system and the inter-turbine grid. The frequency response of the transmission system has been already characterized in previous section, therefore, the next step consist on characterizing the frequency response of the entire electric infrastructure, including the inter-turbine grids. Thus, this second part of the analysis is mainly focused on the inter-turbine grid and its characteristics.

The equivalent impedance of an offshore wind farm varies with changes in the configuration of the inter-turbine grid. As a result, the frequency response of the system varies as well. Thus, the frequency response of the wind farms for different configurations have to be investigated separately [25].

Consequently, based on the transmission system evaluated in section 6.3.2 the analysis performed in this section is focused on the effect of different aspects of the wind farm, like: number of feeders (or radials) in the inter-turbine grid or the location of each wind turbine. The analysis is made from the viewpoint of the wind turbine, which is considered the potential harmonic source in normal operation.

Without the appropriate models is not possible to estimate the resonances of the system. Therefore, the base scenario defined in chapter 5 is used.

This scenario presents a radial design for the inter-turbine grid, where each one of those radials is composed by 6 wind turbines of 5MW. The voltage level of the inter-turbine grid is medium voltage, 33kV. As regards to the spatial disposition of the wind turbines, there is considered as a rectangle (see section 3.3), Figure 6.13.

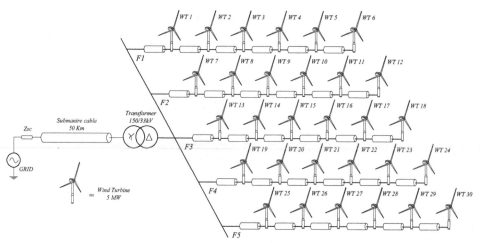

Figure 6.13 The lay-out of the offshore wind farm, which is the base of the resonances analysis.

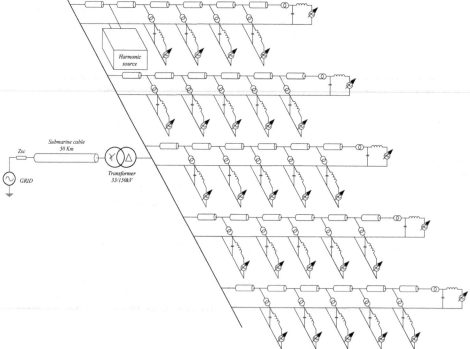

Figure 6.14 The simulation scenario of the offshore wind farm, which is the base for the frequency response analysis.

Considering that the transmission system is equal to the characterized in section 6.3.2, the last feature to define the whole offshore wind farm is the inter-turbine submarine cable. Hence, as inter-turbine submarine cable an ABB XLPE cable with the adequate nominal voltage and power is chosen. The characteristics of this cable are shown in Table 5.21, chapter 5.These characteristics are filled into the PSCAD template, for the model explained in section 4.2.3 with the corrections exposed in section 4.2.4.

The aim of this evaluation is to calculate the frequency response of the entire electric connection infrastructure. Thus, the wind turbine model is not considered as a key issue.

Therefore, the wind turbines are considered as an ideal controlled voltage source with a LCL filter, (see section 5.2.2.3). The filter used to connect the wind turbine to the local inter turbine grid, Figure 6.14.

Taking the scenario depicted in Figure 6.14 as base to estimate the frequency response, the same procedure of section 6.3.2 is used. In this way, to know the frequency response for a specific wind turbine, it is substituted by a harmonic voltage generator (Figure 6.14), which generates a harmonic train in the frequency range of 50Hz-5000Hz, Figure 6.3.

### 6.5.1 Frequency response of a wind turbine depending on its position in the inter-turbine network

Looking at Figure 6.13, it is possible to observe how from the viewpoint of each wind turbine, the equivalent impedance seen is different depending on its location in the inter-turbine grid, i.e. there is not the same equivalent impedance at the output of the 25th wind turbine and at the output of the 30th wind turbine.

To quantify the variation of the frequency responses of each wind turbine, in this section the frequency responses for all the wind turbines of a feeder are estimated. To perform this evaluation, the harmonic voltage source is placed in different positions of the feeder (or radial) and for each position the signals at the PCC are measured. Then, applying the FFT to the signals of the PCC, it is possible to estimate the frequency response for each individual wind turbine.

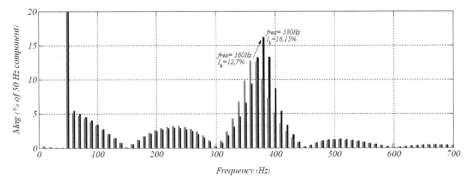

Figure 6.15 Harmonic currents at the PCC depending on the location of the harmonic voltage source inside the inter turbine grid. Substituting the 30th wind turbine (red) and substituting the 25th wind turbine (black).

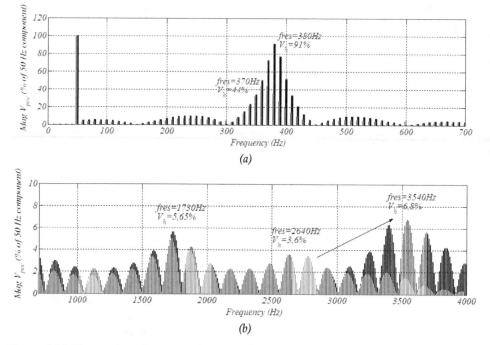

Figure 6.16 Harmonic voltages at the PCC depending on the location of the harmonic voltage source inside the inter turbine grid. Substituting the 25th wind turbine (black) and substituting the 30th wind turbine (red): (a) more detail in the main resonance and (b) more detail in high frequencies.

In the first evaluation, the frequency response of each wind turbine is obtained from the 25th to the 30th. The results for the harmonic currents are depicted in Figure 6.15 and the results for harmonic voltages are depicted in Figure 6.16.

The frequency response of the whole system is similar to the frequency response of the transmission system only. However, the resonance frequency has a short variation. The transmission system presents the resonance at 400 Hz (section 6.3.2, Figure 6.5), but as can be seen from Figure 6.15, the frequency response from the wind turbine viewpoint depends on each wind turbine and is located in the range of 360-380 Hz, close to the 400Hz but not the same.

As seen in section 6.3.3, the step-up transformer does not allow to transmit 3rd order harmonics and multiples. Looking to the results for the 30th wind turbine, the harmonics located at 360Hz and 370Hz have similar amplitudes in both cases (current and voltage), probably the resonance is between them. This fact, can explain the notorious amplitude reduction.

Applying the FFT to the voltage at PCC (Figure 6.16), it is possible to see other two more "small resonances" (attenuated frequencies, but less than the rest), besides the transmission system's resonance.

The first one of these two frequency groups less attenuated than the rest is located between the 1500Hz and 2000Hz. This small resonance has not variations, i.e. is independent to the location of the wind turbine. However, the second group of these frequencies less attenuated depends on the location of the wind turbine. Thus, depending on the location of the wind turbine, the "small resonance" can be around 2500Hz or 3500Hz. The closer is placed the wind turbine to the offshore substation (shorter cable length to the collector point, and less impedance), bigger is the frequency of the resonance.

Note that these two "small resonances" have significantly smaller amplitude than the main resonance at (360Hz-380Hz).

### 6.5.2 Frequency response of the offshore wind farm depending on the feeders in its inter-turbine network

In the first scenario described in section 6.5, 5 feeders (F1-F5) of 6 wind turbines each one are considered, Figure 6.13. However, the internal impedance of the wind farm can have variations with configuration changes, like changes in the number of feeders.

Thus, in this section the frequency response for different inter-turbine grid configurations is evaluated. Inter turbine grids with 2 feeders (F1 and F2) to 5 feeders (F1-F5), Figure 6.13 are considered, maintaining the same number of wind turbines for each radial, not the total number of wind turbines of the wind farm.

In this case, the harmonic voltage source is placed at the first wind turbine of each feeder (7, 13, 19 or 25, Fig. 15), because at this point, the second "small resonance" is closer to the fundamental frequency than in any other location of the feeder, Figure 6.16 (b). The simulation results (FFT of the current and voltage at the PCC) for the considered configurations are depicted in Figure 6.17 and Figure 6.18

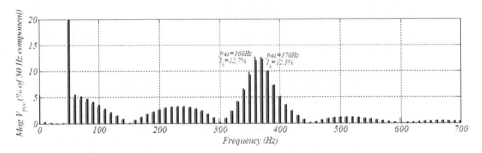

Figure 6.17 Harmonic currents at the PCC depending on the number of feeders in the inter-turbine grid. With 2 feeders F1-F2 (black) and with 5 feeders F1-F5 (red).

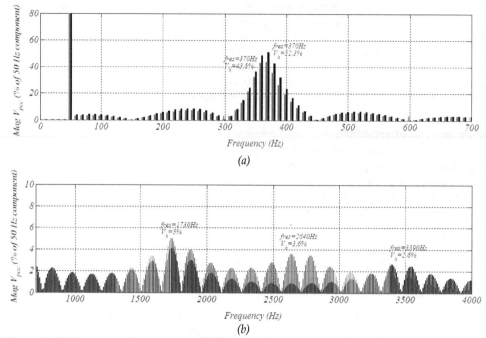

Figure 6.18 Harmonic voltages at the PCC depending on the number of feeders in the inter-turbine grid. With 2 feeders F1-F2 (black) and with 5 feeders F1-F5 (red).

From the evaluation of the results of Figure 6.17, it is possible to determine that the first and main resonance of the system have not big variations for different configurations of the local inter-turbine grid. However, the second of the "small resonances" (frequency groups less attenuated) varies with these configuration changes. If there are fewer feeders, the second "small resonance" occurs at higher frequencies.

### 6.5.3 Frequency response of the offshore wind farm depending on the number of feeders for each step-up transformer's primary

In order to take into account cases where the offshore wind farm have a step-up transformer in the offshore substation with more primary windings than one, the present section analyzes an inter-turbine network configuration with two primary windings, as depicted in Figure 6.19. The purpose of the analysis is to know how affects this extra primary winding to the frequency response of the offshore wind farm.

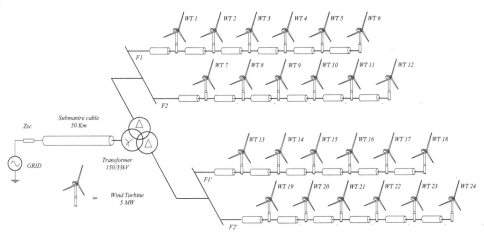

Figure 6.19 Simplified scheme of the simulation scenario of the offshore wind farm with two primary windings.

To know the influence of the extra winding, the results of the configuration depicted in Figure 6.19 and the configuration depicted in Figure 6.13 with only two feeders, are compared. The comparison these frequency responses are served Figure 6.20 And Figure 6.21.

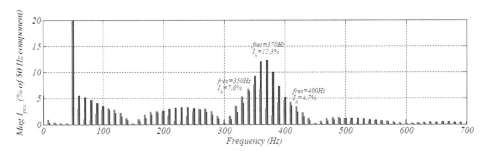

Figure 6.20 Harmonic currents at the PCC. With 2 feeders F1-F2 (red) and with 2 feeders on each primary winding F1-F2 and F1'-F2' (black).

Looking at the results in Figure 6.20, it is possible to see that the use of two windings connected as delta-star(grounded)-delta, where the secondary windings have delta connection, does not allow to transmit multiples of 3rd order divided by two harmonics (multiples of 150Hz/2, 75 Hz) to the PCC. As a result, the frequency response of the system from the viewpoint of the wind turbine presents less harmonic components.

However, as can be seen in Figure 6.21 (b), for the voltage harmonics, there is a new group of frequencies less attenuated at 2520 Hz.

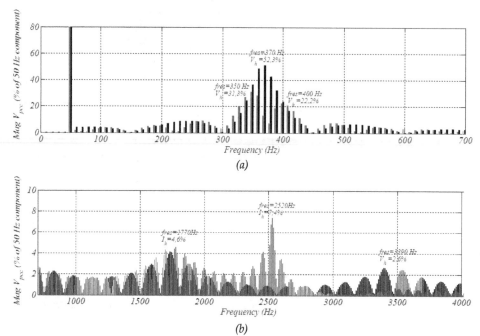

Figure 6.21 Harmonic voltages at the PCC. With 2 feeders F1-F2 (black) and with 2 feeders on each primary winding F1-F2 and F1'-F2' (red).

## 6.6 Harmonic risk evaluation for the considered base scenario

In the previous section, an evaluation of the frequency response of the electric connection infrastructure is carried out. In the next step forward, a specific case for the base scenario considering equivalent wind turbines is analyzed.

The control strategy implemented in the grid side converter of the wind turbine has a notorious influence in the harmonic components injected to the electric connection infrastructure. As is defined in chapter 5, at the grid side converters, a vector control with two current loops is considered, one for the quadratic component and the other for the direct component, in order to control independently both of them. As regards to the modulation, a PWM modulation is used.

As is mentioned before, the simulation of a scenario with 30 complex wind turbine models can be computationally unviable. Thus, starting to the scenario described in chapter 5, a model with an equivalent wind turbine of 30 MW for each radial is developed. As says [110], using N equivalent wind turbines it is possible to have reasonable accurate representation of an offshore wind farm made up with N radials. Thus, the new scenario with wind turbine models is depicted in Figure 6.22.

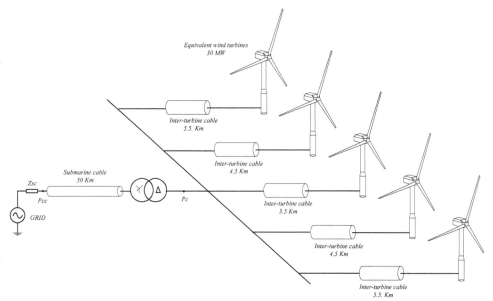

Figure 6.22 Diagram of the simulation scenario with wind turbine models.

To verify that the connection infrastructure of the equivalent simulation scenario has a similar frequency response to the entire offshore wind farm, the frequency response of the equivalent wind farm is obtained via PSCAD. So, using the same procedure of the sections 6.3 and 6.5, an equivalent wind turbine is substituted by a harmonic voltage source (same harmonic train Figure 6.2) and the FFT of the signals is carried out in the PCC, Figure 6.23.

Making out a comparison between both frequency responses, it is possible to verify that the equivalent offshore wind farm and the whole wind farm have a similar frequency response in steady-state.

Thus, comparing the frequency response of the equivalent wind farm (Figure 6.23) with the frequency response of the entire wind farm (Figure 6.15 and Figure 6.16), can be seen how they have roughly the same frequency response.

Finally, the maximum harmonic levels allowed in the main grid for this specific scenario are defined. According to the IEEE-519 standard, for a short circuit impedance of 5% and a 150kV, these limits are as follows, Table 6.8

IEEE-519 Harmonic limits for voltage and current

| $I_{SC}/I_L$ | $n < 11$ | $11 \leq n \leq 17$ | $17 \leq n \leq 23$ | $23 \leq n \leq 35$ | $35 \leq n$ | THD |
|---|---|---|---|---|---|---|
| 20-50 | 3.5 % | 1.75 % | 1.25 % | 0.5 % | 0.25 % | 4.0 % |
| | | | Individual harmonics | | THD | |
| Above 161kV | | | 1.0 % | | 1.5 % | |

Table 6.8 Harmonic limits for voltage and current for the specifications of the simulation scenario.

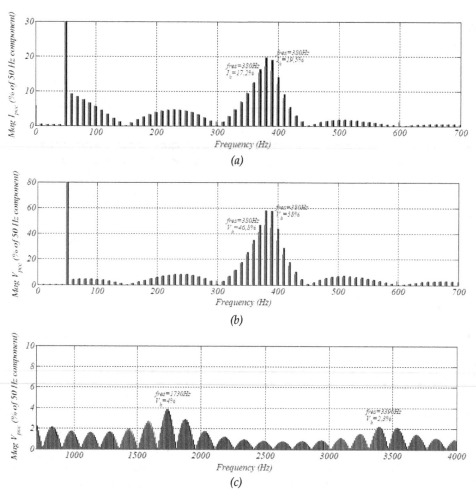

Figure 6.23 Frequency response of the equivalent wind farms model via PSCAD. Frequency response from the viewpoint of the equivalent wind turbine closest the collector point (red) and for the equivalent wind turbine 5.5 Km away the collector point (Blue). FFT of the current (a) and FFT of the voltage (b)-(c).

### 6.6.1 The output voltage of the considered wind turbines

In this section, the harmonic spectrum of the input voltage is characterized, i.e. defined all the harmonics generated by the wind turbines. In this way, the considered input voltage spectrum is based on a PWM modulation and a LC-L filter. The voltage generated in the grid side inverter (referenced to the neuter point) is shown in Figure 6.24.

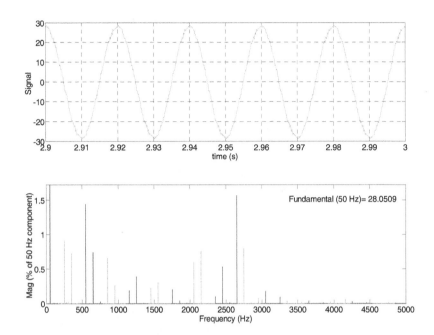

Figure 6.24 Voltage behind the connection filter of the grid side inverter and its frequency spectrum, while the equivalent wind turbine is generating the nominal power.

The voltage with the frequency spectrum off Figure 6.24 is one of the input voltages to the electrical connection infrastructure. For each one of the equivalent wind turbine is considered the same input voltage.

The frequency spectrum depicted in Figure 6.24 is used as an example (like a generic wave of a generic wind turbine) in the base scenario to perform an analysis of the harmonic risk.

So, considering this generic case as an example, it is possible to know how the different harmonics of each equivalent wind turbine interact with the electric connection infrastructure estimating via simulation the harmonic levels at the PCC.

The emitted harmonics for a specific wind turbine varies with the generated active power, at least with the considered control strategy. But, in the present analysis, only the harmonic levels on the PCC for the case while the wind farm is transmitting / generating the nominal power are estimated. This estimation is performed via PSCAD simulation of the scenario that is shown in Figure 6.22, the results are served in Figure 6.25 for current and in Figure 6.26 for voltage.

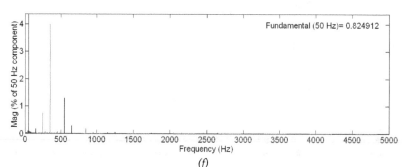

*(f)*

Figure 6.25 FFT of the current at the PCC for the base scenario while is transmitting / generating the nominal active power.

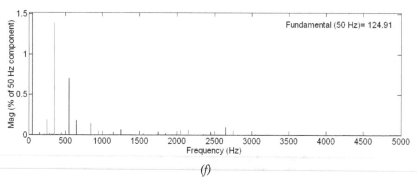

*(f)*

Figure 6.26 FFT of the voltage at the PCC for the base scenario while is transmitting / generating the nominal active power

Carrying out the FFT of the current and voltage at the PCC for the case while the wind farm is transmitting / generating the nominal power, Figure 6.25 and Figure 6.26, can be seen that the harmonic levels are bigger than the limit described in the standard IEEE-519 (Table 6.8), i.e. the harmonic levels at the PCC cannot fulfill the standard IEEE-519. The 350 Hz harmonic (Figure 6.25) is bigger than maximum amplitude allowed for a single harmonic at this point (<3.5%).

From Figure 6.25 and Figure 6.26, it is possible to observe two main harmonics (550Hz and 350 Hz) for current and voltage. The harmonic at 550 Hz is notorious in voltage and current, but this is the resonance frequency of the grid side inverter's connection filter, so it is not amplified by the electric connection infrastructure.

Finally, notice that the harmonic spectrum at the PCC is the combination of the voltage harmonic spectrum of the output voltage of the equivalent wind turbine (Figure 6.24) and the frequency response of the equivalent offshore wind farm (Figure 6.23).

Therefore, the 350 Hz harmonic component is amplified by the energy transmission system of the offshore wind farm. One solution to avoid this amplification caused by the transmission system is the use of resonant passive filters.

Resonant passive filters are based on accept or attract the harmonic current on a specific frequency to the filter branch. At this specific frequency, the RLC branch presents only the resistive part of the impedance, so, at this point the harmonic current is divided depending on the kirchoff law. Thus, the harmonic current is deviated depending on the relation between the impedance of the filter and the impedance of the circuit (see Appendix E: Resonant Passive Filters).

This means that the location of these filters determine the impedance of the circuit in parallel to it. So, the location of the resonant passive filters affects to the current that it is deviated by it.

Moreover, due to the fact that the harmonic currents are deviated to the passive filters, the passive filters do not reduce the harmonic distortion from the harmonic source to the passive filters. Therefore, to avoid the negative effects caused by the harmonics in the electric connection infrastructure (independently of the IEEE-519 standard fulfillment at the PCC) the passive filters have to be placed as close as possible to the harmonic source.

Another option is the use of active filters. Resonant passive filters, as their name describes, are made up exclusively with passive elements to remove a specific harmonic. On the contrary, the active power filters are made up with one or more inverters, usually voltage source inverters (VSI).

Basically, active power filters are made up by: a converter (usually VSI inverter), a device to storage energy (usually a capacitor), several circuits to measure current and voltage and a control circuit to generate the modulated reference signals to the converter. Besides these components, the active power filter can be provided by a transformer or a coil to connect the converter to the grid.

The main advantage of the active power filters in comparison with the passive filters is that they can be adapted to the changing conditions of the load and the electric grid.

Looking at the results displayed in Figure 6.15 and in Figure 6.16, it is possible to see that the main resonance of the system (350Hz) has not huge variations with variations in the configuration of the collector grid. Furthermore, in the present case, there are only two main harmonic frequencies at the PCC (Figure 6.26). So, due to its simplicity, in the present evaluation only passive filters are taken into account.

However, in the wind farm can be implemented many other solutions like active filters or control strategies oriented to mitigate harmonics.

### 6.6.2 Location of the passive filters in the transmission system
As is explained in the previous section, the location of the passive filters determines the impedance of the circuit in parallel to it and the current that it is deviated.

Therefore, in this section advantages and disadvantages to place these filters offshore and onshore are evaluated. Between both locations are some technical differences. For example, offshore side it is possible to connect the passive filter directly to a medium voltage grid, while onshore side only a high voltage grid is available.

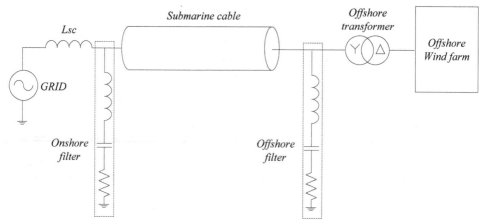

Figure 6.27 Transmission system with onshore and offshore RLC filters.

Nevertheless, in the present analysis, to see in a simple way the difference between these two locations, two identical equivalent filters connected to the same voltage (the transmission voltage 150 kV) are considered, Figure 6.27.

The RLC filters used in this comparative are tuned to filter the main harmonics detected in the PCC of the base scenario: 350 Hz and 550 Hz (Figure 6.25 and Figure 6.26). The characteristics of these filters are shown on Table 6.9.

|  | *Fresonance=350 Hz* | *Fresonance=550 Hz* |
|---|---|---|
| R | 32.45 Ω | 26.3 Ω |
| L | 1.033 H | 0.418 H |
| C | 0.2 μF | 0.2 μF |
| Δf | 5 Hz | 10 Hz |
| Vn (50 Hz) | 150 kV* | 150 kV* |
| In (fo) | 53.4 A | 25 A |
| *Equivalent filters transformed to this voltage* | | |

Table 6.9 Characteristics of the considered RLC filters.

Therefore, based on the scenario described in the previous section, the analysis of the best location for passive filters is carried out via PSCAD simulation. For that purpose, the passive filters with the characteristics of Table 6.9 are placed onshore and offshore of the equivalent offshore wind farm, Figure 6.22.

These filters have the objective to reduce the harmonic levels at PCC, so, in order to compare the simulation results and deduce conclusions, in both cases (for offshore filters and onshore filters), the FFT of the current is performed at this point.

These simulation results are depicted in Table 6.10, in this table, three results are compared: results of the base scenario with onshore filters, results of the base scenario with offshore filters and results of the base scenario without filters. These results are for the case, when the transmission system is transmitting the nominal power, Figure 6.25.

|                | $I_{350Hz}$ | $I_{550Hz}$ | THD  |
|----------------|-------------|-------------|------|
| Without filter | 4           | 1.3         | 4.52 |
| Onshore filter | 1.96        | 1.15        | 2.43 |
| Offshore filter| 1.62        | 1.11        | 2.09 |

Table 6.10 Comparison of the amplitude of the harmonic currents in the PCC for different positions of the filters, while the wind farm is generating the nominal power.

Looking at Table 6.10 seems that the best location to place RLC filters is offshore, due to the fact that the THD at the PCC is smaller in comparison with the other options. Furthermore, with the passive filters placed offshore, it is possible to see a bigger reduction in all the selected individual harmonics. Besides, in offshore location there is the possibility to connect these filters to the medium voltage inter-turbine grid.

However, offshore location also has disadvantages, the access to the location is more difficult and the passive filters need a place to be located in the offshore platform.

Last of all, note that the passive filters definition carried out in the present analysis is based on the perfect knowledge of parameters, therefore, in real conditions where the parameters are not well known, passive solutions might not be suitable.

## 6.7 Chapter conclusions

The study presented in this chapter is focused on the evaluation of the frequency response of the offshore wind farm. This frequency response depends on design parameters such as: the cable length and characteristics, transformers connection and leakage inductance or inter-turbine grid's configuration. The analysis carried out estimates the potential risks on the voltage and current harmonic amplifications.

For that purpose, the state equations are a good approximation in order to estimate in an easy way the frequency response and main resonances of the system. The results obtained with this method are very similar to the simulation results in PSCAD.

As regards to the harmonic risk of the AC offshore wind farms, this kind of wind farms have the potential to amplify low order harmonics due to the iteration between the capacitive component of the submarine cable and the leakage inductance of the step-up transformer. From the results of this study, it is possible to observe, that the resonance frequency depends mainly on the characteristics of the submarine cable (its capacitive component).

The main resonance of the AC offshore wind farm from the viewpoint of the wind turbines is the same of the transmission systems resonance. The inter turbine grid, does not cause big variations in the frequency response and for different positions in the inter turbine grid, the frequency response is similar. However, the inter turbine grid causes "small resonances", which varies with the wind turbines position in the inter-turbine grid. This little resonance has less potential to amplify harmonic components, but, grid codes (like IEEE-519 standard) are more restrictive with the high order harmonics.

To avoid as far as possible the harmonic amplification in normal operation due to the resonance of the transmission system, one good option seems to choose a configuration which the resonance frequency of the transmission system coincides with one of the frequencies that the step up transformer does not allow to transmit, Figure 6.7.

Finally, highlight that the present analysis does not takes into account the effect of the control strategies of the wind turbines. These control strategies can be oriented to mitigate the resonance avoiding filters. In the same way, neither is taken into account further and more complex analysis about the harmonic risk or problems related to the iteration between the control strategies of the wind turbines.

# Chapter 7

# Analysis of Disturbances in the Power Electric System

The analysis performed in the previous chapters of this report is focused on normal operation and steady-state. Nevertheless, as any other installation, offshore wind farms must be protected against different eventualities.

For instance, when lightning hits the ground in the vicinity of a high-voltage transmission line or when lightning strikes a substation directly, the grid changes from one steady-state to another and a transient occurs. However, the majority of power system transients are the result of switching actions which are normally required for the ordinary operation of the electrical network.

Grid codes have specific requirements for voltage dips or faults at the PCC. So, these types of disturbances are well defined. Inside the offshore wind farm is also possible to appear disturbances, such as faults or disturbances related to breakers operation. Nevertheless, unlike for those ones at the PCC, for disturbances inside the wind farm are not specific requirements only protect the system. To this end, must be avoid instabilities and dangerous voltage / current peaks in the system.

Therefore, the normal operation analysis is not enough to carry out the pre-design of an offshore wind farms electric system. So, in the present chapter those disturbances and their associated transients are evaluated.

## 7.1 Transients in electric power systems

If a change in the normal conditions takes place in the power electric system, a temporary transient occurs. Consequently, due to the inherent dynamics of the electric system, the system needs a period of time to re-establish its previously steady-state condition.

A change in the normal conditions of the system can be programmed or accidental. If a lightning hits the ground close to a high-voltage transmission line or strikes directly a substation, the grid changes from one steady-state to another. As a result, this event will cause a transient.

Aside from lightning strikes, breaking actions also changes the system from one steady-state to another. Furthermore, the most of the power transients are caused by switching actions related to the breakers operation which, connects / disconnects parts of the network under load and no-load conditions [113].

The transients caused by breaking actions are related to the nature of the interrupted current. Thus, the transient after a disconnection during normal operation or after the clearing of a fault are considerably different. In this way, because of its magnitude and severity, short circuit current interruptions are especially of concern.

When a fault, in the form of a short circuit current, occurs in an electrical system and as a result of such fault, the part of the electric system where is localized the fault is disconnected. The protection device switches and interrupts the short-circuit current.

When a short-circuit current is interrupted, even at current zero as do the current interrupting devices, the magnetic energy stored in the leakage inductance of the transformer at the substation, in the inductance of the connection bus bars or in the submarine cables is still there.

Taking into account only a short period of time, it is possible to approximate a short-circuit current during a fault as a steady-state situation where the energy of the system is mainly stored in the magnetic field (high currents). So, when the fault current is interrupted, the system changes to another situation, where without current, the energy is predominantly in the electric field.

Between those situations, the magnetic energy stored in the mentioned components is transferred from the magnetic field to the electric field to adequate to the new scenario. As a result of this energy exchange a voltage oscillation appears in the system.

At a fault clearance, the fault and the disconnection (current interruption), both events are destabilizing changes to the system. The interruption itself produces an additional transient superimposed upon the instantaneous conditions of the system. Thus, interrupting devices must deal with transients in the currents generated elsewhere (for example due to a fault) and voltage transients caused by the interrupting device itself.

The voltage response at the instant of current interruption and the first few microseconds thereafter mainly depends on the lumped elements. After the first few microseconds, travelling waves have an important role in the transient recovery waveform.

The electromagnetic wave, which is caused by the interrupting device, propagates along a transmission line (constant characteristic impedance) with a fixed relation between the voltage and current waves. But if the wave arrives at a discontinuity, such as: an open circuit, a short circuit or a point where the characteristic impedance changes, an adjustment of the voltage and current waves must occur.

At the discontinuity, a part of the energy is let through and a part of the energy is reflected and travels back. In the case where the losses are neglected, the total amount of energy in the electromagnetic wave remains constant. Therefore, electromagnetic waves propagate through the system even after current interruption. Consequently, the resistive part of the circuit determines the losses of the electromagnetic wave and the duration of the oscillation caused by the switching device itself.

In this way, the accumulation of reflected electromagnetic waves with local oscillations gives the voltage waveform at the terminals of the interruption device. So, the shape of this waveform depends on the interrupted current, cable lengths, the propagation velocity of the electromagnetic waves, the reflection rates at the discontinuities of the system and the configuration of the power electric system.

In resonant scenarios with the potential to amplify some harmonics, the electromagnetic wave caused by the switching action can excite the resonant frequency of the system, which can causes additional over-voltages and instabilities to the system.

In short, switching operations and specially fault current interruptions have to be into account to avoid damages in the system caused by transient currents and voltages. Even more in systems such as the inter-turbine grid of the offshore wind farm with a resonant amplification potential. As a result a study of the transient is needed.

### 7.1.1 Fault clearance in the inter turbine grid

After the fault clearance, when a short circuit current is interrupted (even at zero current), a transient recovery voltage (TRV) will appear across the terminals of the interrupting device. The configuration of the network as seen from the terminals of the switching device determines amplitude, frequency, and shape of the current and voltage oscillations [114].

To determine the shape of the transient recovery voltage across the terminals of the circuit breaker, while clearing short circuits, many investigations are carried out. Thus, for different types of grids, a typical shape for the TRV is determined.

For large substations, like the considered offshore substation, if a fault occurs in one of the feeders, due to the number of adjacent feeders in parallel, the characteristic impedance seen by the circuit breaker is considerably smaller than the characteristic impedance of the faulted line. Therefore, for cases with the following network operation, Figure 7.1, the TRV exhibits the shape shown in Figure 7.2, [115] – [116].

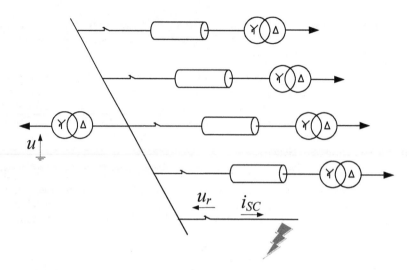

Figure 7.1 Generic large substation composed by several feeders.

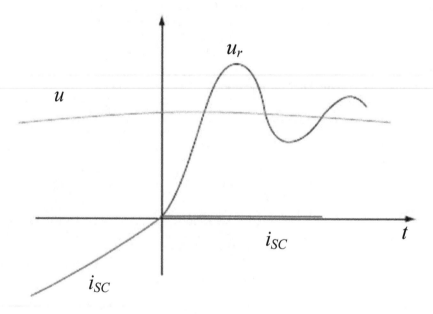

Figure 7.2 TRV for the considered network.

So, the considered inter-turbine grid for the base scenario has the potential to cause over voltages and oscillations at the clearance of a short circuit current in the inter-turbine grid.

### 7.1.2 Submarine cable energizing / de-energizing

Voltage and current transients can appear even earlier than faults, when a circuit is being energized or de-energized. In this way, the energizing / de-energizing transients of transformers, reactors, and submarine cables are especially of concern [117], [118].

The simple closing of a switch or of a circuit breaker to energize a circuit can produce significant over-voltages in an electric system. These over-voltages are due to the system adjusting itself to an emerging different configuration of components.

In the first step, a simply generic case is analyzed, a cable that is being energized through a transformer. For the sake of simplicity, a capacitor C is used to represent the transmission cable. The rest of the circuit is represented by an equivalent inductance and resistor. As a result of this simplification, the equivalent circuit can take the form of the circuit illustrated in Figure 7.3 (a second order equivalent circuit).

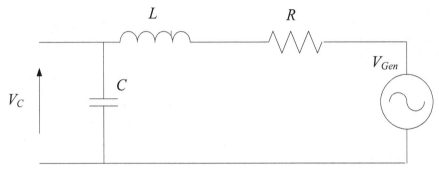

Figure 7.3 Single-phase RLC circuit.

Energizing a capacitive load is a well-known transient event, causing transient over voltages and inrush currents depending on the amplitude of the input voltage and the resistive part of the circuit. The equations in "s" domain that describe the behavior of the RLC circuit are:

$$\frac{V_C(s)}{V_{Gen}(s)} = \frac{1}{LCs^2 + 2RCs + 1} = \frac{1/LC}{s^2 + Rs/L + 1/LC} \equiv \frac{\omega_n^2}{s^2 + 2\xi\omega_n s + \omega_n^2} \qquad (141)$$

$$2\xi\omega_n = \frac{R}{L} \qquad (142)$$

$$\omega_n^2 = \sqrt{1/LC} \qquad (143)$$

As it can be seen in the equations (141) - (143), the transient response is determined in amplitude by the inductor combined with the resistive part. As regards to the oscillation frequency, is determined by the relation between the inductor and the capacitor.

The voltage on the capacitor during the energizing transient of the RLC circuit is depicted in Figure 7.4

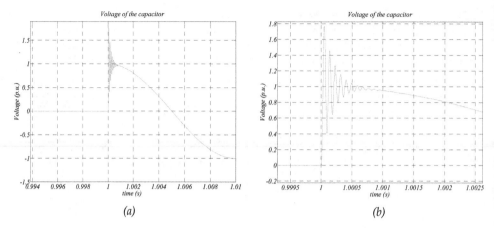

*(a)*                                                    *(b)*

Figure 7.4 Transient voltage of the capacitor during the energizing the RLC circuit, (a) Voltage on the capacitor and (b) voltage on the capacitor with more detail around the energizing.

The transient voltage, shown in Figure 7.4, oscillates along the line at a relatively low frequency (1/LC) for dumping factors lower than one ($\xi$ < 1). It has an amplitude which can reaches a peak value approximately equal to twice the value of the system voltage that was present at the instant at which the closure of the circuit took place (if $\xi$ = 0).

With regards to the current at the energizing / de-energizing, the series impedance limits the inrush current.

The case depicted in Figure 7.4, only shows the energizing of the capacitor, but the equations (141) - (143) also describes the behavior of the circuit during the de-energizing, because, these equations are calculated to describe the relation between the input voltage ($V_{Gen}$) and the voltage of the capacitor ($V_C$).

Applying a short circuit across the capacitor in this circuit is equivalent to applying a line to ground fault on a single-phase power system [117]. Thus, considering that the cable has a little resistive component ($\xi$ < 1) and unless the cable energizing, a short circuit of 80% of depth. This simple circuit will behave as in Figure 7.5 [113], [115] and [117].

Note that the results illustrated at Figure 7.5 are a superposition of an AC voltage reduction (80%) and the transient depicted in Figure 7.4. The amplitude of the transient depends on the instant that the circuit breaker is closed (amplitude of the voltage variation) and the resistive part of the circuit.

With regards to the simplification of the system in lumped parameters, as is explained in chapter 4, section 4.2.2., it is possible to use of lumped parameters to represent a electric circuit, if the physical dimensions of the power system, or a part of it, are small compared with the wavelength of the voltage and current signals.

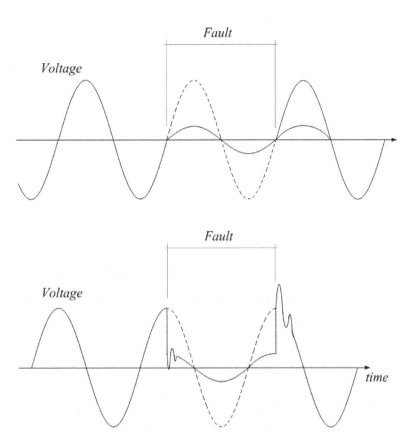

Figure 7.5 Voltage transients for a single phase RLC circuit, (a) the capacitor is short circuited at zero voltage and (b) the capacitor is short circuited at the maximum voltage.

Finally, to extrapolate the results of the single-phase system, a three-phase power system can be treated as a single-phase system when the loads, voltages, and currents are balanced.

The results of this simple system, considering a balanced three-phase voltage, are depicted in Figure 7.6.

One of the main differences between Figure 7.3 and the transmission system of the offshore wind farm is the inductive shunt compensators. As is explained and defined in chapter 4, the shunt reactors are used to improve the energy transfer capability of the submarine cable, reduce the active power losses, etc...

Therefore, the considered transmission system presents shunt (parallel) reactors at both ends of the line. As regards to how affects those reactors to the energizing of the submarine cable, [118] says that the shunt inductive compensator reduces over-voltages at the cable energizing. Thus, it recommends the connection of those reactors before the energizing of the cable to prevent over-voltages.

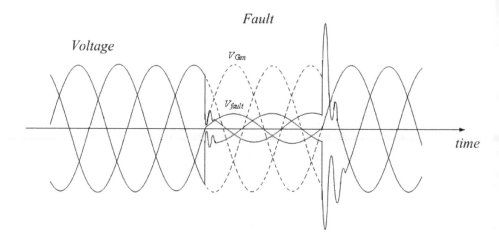

Figure 7.6 Voltage transients for a three phase fault balanced in a RLC circuit.

In short, the disturbances associated to the energizing / de-energizing of the submarine cables such as voltage dips at the PCC, have the potential to provoke over voltages and oscillations in the offshore wind farm.

Due to this fact, in the following sections, these disturbances which can cause instabilities and current / voltage peaks are evaluated upon the base scenario.

## 7.2 Considered scenario for the problem assessment

In order to perform grid studies, often complete models are not suitable because of the high computational effort required. Therefore, for such kind of analysis, usually reduced models are used [110], [119].

For this reason, in the present section, starting on the scenario with 30 complex wind turbine models described in chapter 5, an equivalent model is developed. As a result, the entire offshore wind farm can be simulated within acceptable time frames.

In contrast with the scenario of the previous chapter, this scenario is focused on evaluate the transient response of the electric connection infrastructure. Consequently, in the new simulation scenario, in order to model the behavior of the offshore wind farm, the main electronic aspects of the wind turbines are considered, like: the power electronic devices and their associated control strategies.

To this end, the scenario depicted in Figure 5.9, is simplified to another scenario with less computational cost. This simplification is carried out by using equivalent wind turbines instead of the complete feeders, i.e. the feeders composed by six wind turbines are substituted by an equivalent one. As claimed in [110] for instance, using N equivalent wind turbines it is possible to have reasonable accurate representation of an offshore wind farm made up with N radials. This simplification is validated in Appendix F: Comparison and validation of the equivalent feeder.

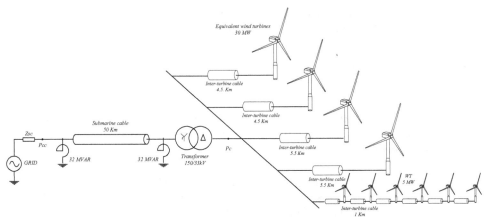

Figure 7.7 Considered scenario for the problem assessment.

Nevertheless, in order to analyze disconnection of wind turbines, one of the feeders of the wind farm is not simplified. Thus, the considered base scenario for the problem assessment is shown in Figure 7.7.

To simplify the real feeder (composed by six wind turbines, section 5.2.2) into an equivalent wind turbine, ideal voltage sources are used, i.e. the six converters of a feeder are simplified into a controlled voltage source.

Upon this voltage source an equivalent control is implemented to achieve the same behavior of the equivalent feeder and the full feeder. In other words, the ideal voltage source is provided with a control strategy with equivalent dynamics and behavior of the real feeder.

In the same way, the filter inductance is sized to keep the same dynamics and wave quality for the equivalent feeder and the full feeder, Figure 7.8.

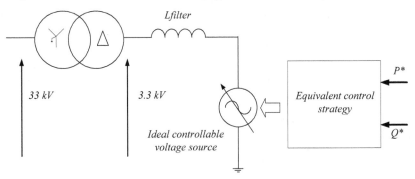

Figure 7.8 Control and power circuit of one phase of the equivalent wind turbine.

As regards to the inter turbine cables, those are simplified in an equivalent length. This equivalent length is calculated depending on the current though the inter turbine cable [120].

Through the cable segment used to connect the wind turbine N°29 and wind turbine N°30 only flows the energy generated in the wind turbine N° 30 (Figure 6.13). Nevertheless, through the cable segment between the PC and the wind turbine N°25 flows all the energy generated in the whole feeder. So, the equivalent length of the wind turbine cable is calculated as follows.

$$l_{equi} = \frac{6 \cdot S_{pc \to 25} + 5 \cdot S_{25 \to 26} + 4 \cdot S_{26 \to 27} + 3 \cdot S_{27 \to 28} + 2 \cdot S_{28 \to 29} + S_{29 \to 30}}{6} \quad (144)$$

Where: $l_{equi}$ is the equivalent inter-turbine cable length, and $S_{pc \to 25}$ is the cable segment length between the collector point and the wind turbine N° 25 (1 Km) and $S_{NN \to NN}$ inter turbine cable segments between wind turbines (1 Km).

### 7.3 Evaluation of breaker operations in the inter-turbine grid

A breaker operation can provoke instability on the system or voltage / current peaks as is discussed in section 7.1. Thus, in order to prevent problems associated to these actions. In the present section these kinds of operations are evaluated.

Based on the complete electrical scheme described in chapter 5 (Figure 5.25 in which are detailed the breakers and fuses), the points of the electric system where the wind turbine and feeder breaking operations are allowed are defined.

The protection system has breakers in each wind turbine after the step up transformer and for each feeder in the offshore platform.

As is mentioned in section 7.1, the power transients are caused by breaking actions under load and no-load conditions. Due to the fact that the transient caused by the breaking action are more severe depending of the magnitude of the interrupted current, only the most severe switching operations are evaluated: on-load switching operations. Therefore, in the present section three different cases are evaluated:

- The disconnection of one wind turbine from its breaker while this is operating at full-load (at zero crossing of the current).
- The disconnection of a feeder while this is operating at full-load (at zero crossing of the current)
- The disconnection of a faulted feeder (at zero crossing of the current).

As regards to the causes which can force those breaking operations, they can be several. For instance, the disconnection of a wind turbine can be forced due to a malfunction in the power electronics.

In the same way, if due to a fault, a short circuit current is circulating through one feeder, the faulted feeder must be disconnected, i.e. the breaker associated to this feeder (BRK 3 to BRK 7, Figure 5.25) has to interrupt the short circuit current.

### 7.3.1 On load breaker operations: Disconnection of wind turbines

The first of the considered switching action to analyze are the on-load wind turbine disconnections. In this way, if this disturbance causes instabilities in the system or unacceptable current or voltage peaks is evaluated.

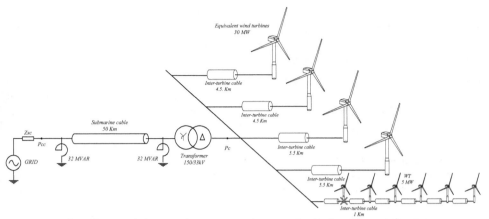

Figure 7.9 The scheme of the simulation scenario to evaluate the effect of the disconnection of a wind turbine.

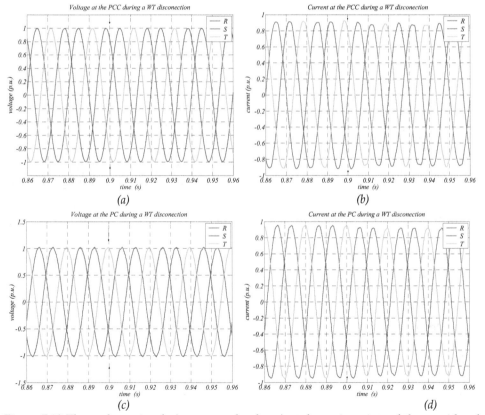

Figure 7.10 Three-phase signals (current and voltage) at the main points of the considered wind farm when the disconnection of a wind turbine occurs.

Thus, a disconnection of a wind turbine is simulated while this is generating the rated power. The simulation scheme is depicted in Figure 7.9 and the simulation results are shown at Figure 7.10. The disconnection order is given to the breaker at 0.9s (the breaker interrupts the current at zero crossing instant).

This disconnection does not cause a huge current reduction to the system. As a result, the transient does not cause any instability or peak problem at the collector point or at the point of common coupling. Furthermore, looking to the results illustrated in Figure 7.10, it is not possible to see any voltage or current oscillation caused by the disconnection.

### 7.3.2 On-load breaker operations: Disconnection of feeders

The second considered on-load switching operation is the disconnection of a feeder. As for the wind turbine disconnection, for the feeder disconnection also the wind farm and the disconnected feeder are generating the rated power. The disconnection order is given to the breaker at 0.9s and the breaker interrupts the current at zero crossing instant. The simulation scenario is shown in Figure 7.11 and the simulation results are depicted at Figure 7.12.

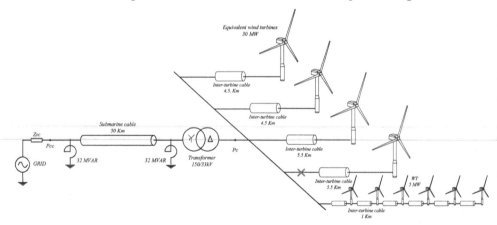

Figure 7.11 The scheme of the simulation scenario to evaluate the effect of the disconnection of a feeder.

The disconnection of a feeder does not cause a significant current reduction to the system, thus, the transient caused by the circuit open does not provoke any instability or peak problem at the collector point or at the point of common coupling. Nevertheless, it is possible to observe a signal oscillation at the PC, Figure 7.12(c).

The protection breaker to disconnect a feeder is suited in the offshore substation, which means that if the feeder is disconnected, the resultant electric scheme is different in comparison with the scenario before the disconnection. It changes the capacitive component so the oscillation frequency does too (see chapter 6, section 6.5.2).

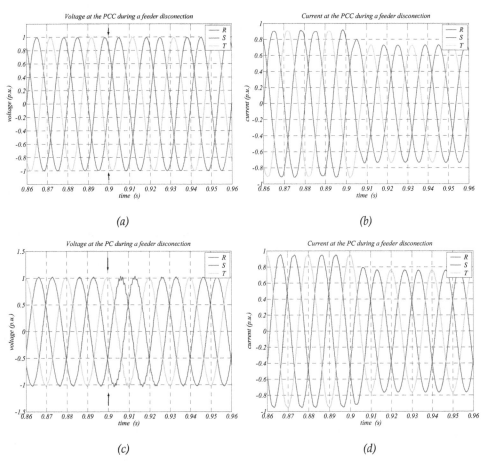

Figure 7.12 Three-phase signals (current and voltage) at the main points of the considered wind farm when the disconnection of a feeder occurs.

### 7.3.3 On-load breaker operations: Fault clearance in the inter-turbine grid

For the last case, the disconnection of a feeder at zero current during a fault (phase to phase) is simulated. A fault causes a short circuit current through the feeder associated to BRK7. So, to protect the integrity of the system, after 60ms (0.9s) of short circuit (the delay time considered to operate the breaker), the breaker interrupts the current at zero crossing instant. The simulation scenario is shown in Figure 7.13.

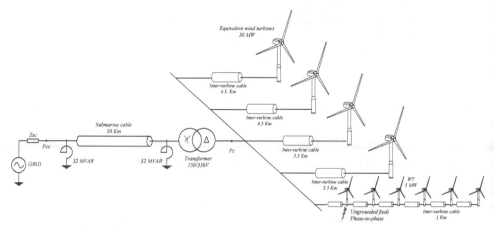

Figure 7.13 The scheme of the simulation scenario used to evaluate the effect of the disconnection of a feeder during a short circuit fault (phase to phase).

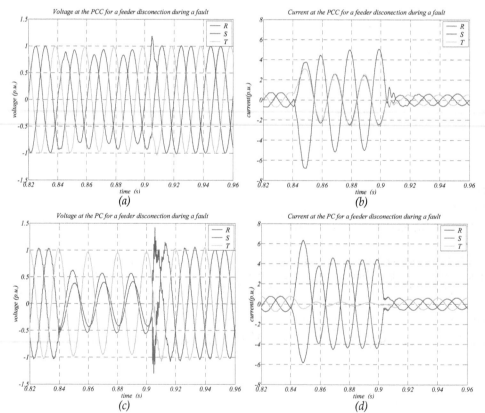

Figure 7.14 Three-phase signals (current and voltage) at the main points of the considered wind farm when a feeder is disconnected during a short circuit fault.

In the results depicted in Figure 7.14, it can be observed that the short circuit current is given mainly by the main grid, due to the fact that the wind turbines have a current limit and they cannot inject infinite current. This causes an unbalanced transient in the PCC, but this transient has a short duration (60ms) and does not causes instabilities to the system.

Looking at Figure 7.14 (c), there is a voltage peak at the PC with a transient oscillation (naturally dumped). The voltage peak has the amplitude close to the 50% over the nominal voltage. As regards to the frequency of the transient oscillation, this frequency is 1200 Hz, the natural frequency of the system.

These voltage oscillations with a peak value of 50% of the nominal voltage, can damage the power electronic devices of the system. However, the voltage and current peaks measured in simulation are not dangerous for the main elements of the transmission system: Step-up transformer, submarine cable and inductive compensators. Although, according to some studies, high frequency transients (not only peak value) are apparently the responsible of insulation failures in transformers [121].

In this way, in Middelgrunden and Horns rev wind farms almost all the transformers had to be replaced with new ones due to insulation failures [122] and in [123] it is suspected that the fast switching breakers caused the insulation failures in these transformers.

Nevertheless, the direct proof of the negative impact of the high frequency transients on the transformer insulation is not yet found [124].

Over-voltages in transmission and distribution systems cannot be totally avoided; however, their effects can be minimized, i.e. the magnitude of the over-voltage can be limited by the use of appropriate elements.

Nevertheless, the objective of this book is the better knowledge of the nature of those transient over voltages and currents, i.e. identify the possible problematic aspects. So, the mitigation of the voltage peaks by using protection devices (surge arrestors, current limiting inductors, etc…) is not analyzed.

However, there is considered the use of passive resonant filters as blocking filters to modify the frequency response of the system, because, in this way, it is possible to evaluate the transient oscillations phenomena. Due to the fact that these filters are oriented to change the frequency response of the system, in order to avoid the transient oscillations and over-voltages.

### 7.3.3.1 Fault clearance in the inter-turbine grid with passive filters

When a circuit breaker interrupts a current, even at zero current, it causes a transient. This transient is caused by the system adjusting itself to an emerging different configuration of components, as is explained in section 7.1.

In circuits with inductive-capacitive components, the transient has oscillations. The energy exchange between the inductive component and the capacitive component of the circuit produces an oscillation at the resonance frequency.

A simple method to mitigate those oscillations and over voltages is to install a blocking filter in series with the generator and the step-up transformer winding or connected to the neutral point of the high voltage side of the transformer [125].

Thus, if a RLC passive filter is adjusted at this resonance frequency, the oscillating energy between those components is deviated to the filter to be consumed in its resistive part.

For the considered base scenario, the oscillation is caused mainly by the iteration between the leakage inductance of the step up transformer in the offshore platform and the capacitive component of the inter-turbine cables. The inter turbine cables are connected in parallel to the collector point, thus, the total capacitive component of the inter turbine grid, is the addition of the capacitive components of all the inter turbine cable segments. As a result, with more capacitive impedance in parallel, the frequency of the resonance decreases.

The disconnection of a complete feeder from the breaker suited in the offshore platform, changes the morphology of the system, as well as the equivalent impedance of the circuit. So, the number of feeders connected to the PC determines the voltage peak and the oscillation frequency of the transient caused by the disconnection of a feeder.

Therefore, in this section, the frequency response variations due to the changes in the inter turbine grid are analyzed. More specifically, the frequency response variation changing the considered wind farm from five feeders connected to only one feeder connected.

To perform the harmonic analysis of the system, in the same way of the previous chapter 6, in this chapter too, the frequency response of the equivalent scenario is estimated using a harmonic source. Furthermore, also the same harmonics train of the previous chapter (Figure 6.2) is used.

In the present case, the scenario to estimate the frequency response is shown in Figure 7.15.

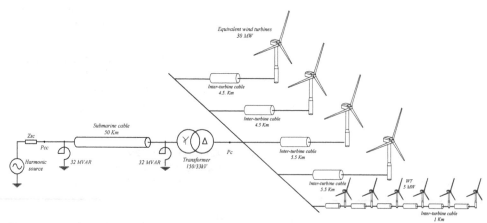

Figure 7.15 The scheme to estimate the frequency response of the considered scenario to evaluate disconnections of wind turbines and feeders.

The results for the scenario displayed in Figure 7.15, varying the number of feeders connected to the PC are summarized in Figure 7.16.

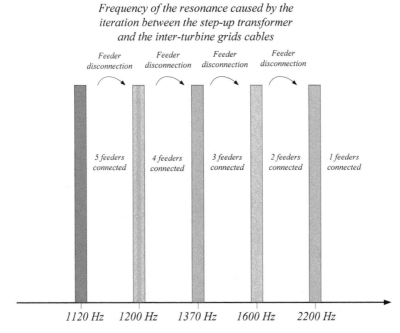

Figure 7.16 Frequency of the resonance caused by the iteration between the step-up transformer and the inter turbine cables depending on the number of feeders connected to the PC.

A wind turbine disconnection does not change the electrical structure of the system as much as a disconnection of a feeder. In case of wind turbine disconnections, parallel impedance is disconnected from the electric infrastructure. Therefore, the system has not huge variation, still has a similar frequency response, as can be seen at the results summarized in Figure 7.17.

The objective is to change the frequency response of the system, to eliminate or attenuate the resonances and avoid the harmonic amplifications and instabilities to the system. For that purpose, for each one of the frequencies with the potential to amplify harmonics must be placed a passive filter. Thus, the passive filters are tuned to filter the resonance frequencies of the complete offshore wind farm and for its possible variations.

In this way, passive filters to avoid resonances of the system are placed for each possible combination, i.e. a passive filter adjusted at 1200Hz is placed to avoid the oscillation caused for the resonance of the system when the wind farm is operating with four feeders connected, another filter adjusted at 1370Hz is placed to avoid the oscillation caused for the resonance of the system when the wind farm is operating with three feeders connected, etc…In short, five passive filters are placed to avoid the resonances for all the possible cases and combinations of the offshore wind farm.

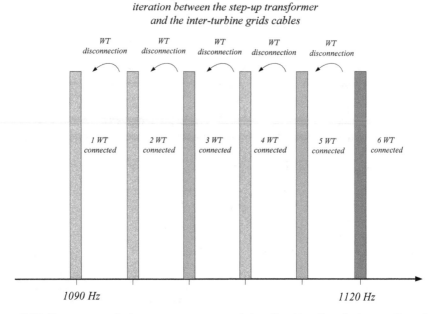

Figure 7.17 Frequency of the resonance caused by the iteration between the step-up transformer and the inter turbine cables depending on the number wind turbines connected to the feeder.

The passive filters are adjusted based on the following characteristics: reactive power generated at 50 Hz 3 MVAR, rated current of the RLC branch 1000A (Imax at resonance frequency). To know how are adjusted the passive filters see Appendix E: Resonant Passive Filters. The characteristics of the used passive filters are shown in Table 7.1

| Fresonance | R | L1 | L2 | C | | Ploss (50Hz) |
|---|---|---|---|---|---|---|
| 1100 Hz | 19 Ω | 1.65 mH | 7.95 mH | 8.3765 μF | 2.5 kW | 0.0016 p.u |
| 1200 Hz | 19 Ω | 1.37 mH | 7.95 mH | 8.3765 μF | 2.5 kW | 0.0016 p.u |
| 1370 Hz | 19 Ω | 0.927 mH | 6.36 mH | 8.3765 μF | 2.5 kW | 0.0016 p.u |
| 1600 Hz | 19 Ω | 0.665 mH | 6.36 mH | 8.3765 μF | 2.5 kW | 0.001 p.u |
| 2200 Hz | 19 Ω | 0.259 mH | 4.77 mH | 8.3765 μF | 2.5 kW | 0.0006 p.u |

Table 7.1 Characteristics of the passive filters used to improve the transients during disconnections.

The passive filters, as well as the inter-turbine cable generate capacitive reactive power. If this reactive power is not compensated, the step-up transformer has to be oversized. Therefore, depending on the number of passive filters and the reactive power generated by them, an inductive static reactive power compensator has to be placed.

In this way, in the present section, the simulations are carried out taking into account the five passive filters and their associated reactive power compensation, Figure 7.18.

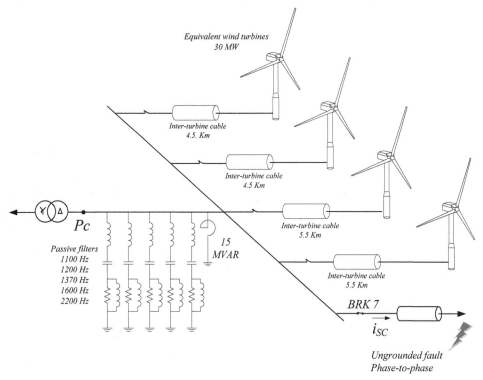

Figure 7.18 The scheme of the simulation scenario to evaluate the effect of the disconnection of a feeder during a short circuit fault.

To evaluate if the passive filters improve the transient response of the system, the disconnection of a feeder during a fault is simulated. As in the previous section, after 60ms (3 cycles) of short circuit (0.9s), the breaker interrupts the current at zero crossing instant. The simulation results of the base scenario with passive filters for this first case are depicted at Figure 7.19.

Therefore, from the simulation results displayed on this section, it can be concluded that the use of passive filters adjusted at the resonances of the offshore wind farm avoids the high frequency transients and reduce current / voltage peaks at breaker open operations.

The reduction of the high frequency components of the transient have a considerable importance, because as is mentioned before, according to some studies the high frequency transients are the cause of the transformer insulation failures [121]. Although, the direct proof of the negative impact of the high frequency transients on the transformer insulation is not yet found [124].

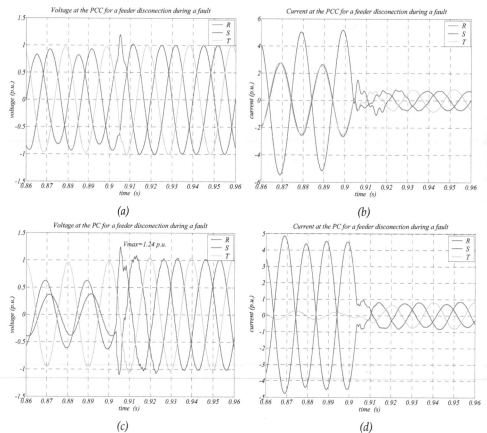

Figure 7.19 Three-phase signals (current and voltage) at the main points of the considered wind farm with passive filters when the disconnection of a feeder occurs during a fault.

## 7.4 Evaluation of the voltage dips in the PCC (LVRT)

The term LVRT is used to describe the capability of an electric system (in this case the offshore wind farm) to get over the voltage dips.

A voltage dip is a sharp voltage reduction down to 90-10% of the RMS voltage magnitude of one phase (or several) with a quick recuperation of the nominal values. So, a voltage dip can be described using two parameters: the depth and the duration.

Focusing on wind power based generation systems, the behavior of these types of generation plants during any eventuality, which causes a voltage drop at the point of common coupling (PCC), is crucial to ensure the quality and the continuity of the electric supply.

If due to a voltage disturbance in the grid, like a voltage dip, all the wind generation systems have been disconnected from the grid. After the clearance of the fault, there will not be enough active power to supply the consumption, which can cause the collapse of the distribution grid.

Thus, in order to avoid the collapse of the distribution system, the wind generation systems have to be provided with the capabilities to remain connected to the grid during voltage dips in the PCC. Furthermore, wind generation systems have to help the distribution system during the faults to recover the normal operation.

Therefore, if this kind of generation systems wants to be connected to the distribution grid, they have to satisfy the grid code requirements of the system operator (SO).

Transmission system operators often put strict requirements on the low voltage ride through (LVRT) of wind power plants. The specific details of the requirements differ significantly between transmission system operators. However, it is possible to find in them some common features on voltage ride trough. Roughly, the behavior required to wind farms by international grid codes for voltage dips in the PCC has the following common specifications

- The tripping of the energy generation resource as a result of these short-term faults should be prevented, i.e. all the wind turbines of the wind farm must remain connected to the grid during any voltage dip (at the PCC) within the boundaries displayed at Figure 7.20.
- During the fault, the wind farm must support the grid voltage increasing the injected capacitive reactive current to the PCC.
- After the clearance of the fault, the wind farm must recovery the injected active power to the values before the fault fast and in an orderly way. In other words, the wind power plant production should recover to its pre-fault value within a certain time after the clearing.

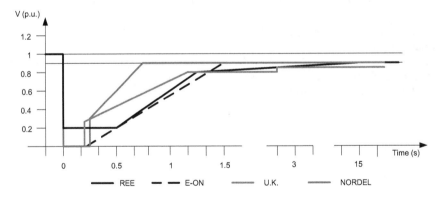

Figure 7.20 Profile of the voltage dips that a wind turbine must remain connected for several grid codes.

In order to focus the analysis, the REE grid code (see Appendix C: REE Grid Code Requirements for Voltage Dips) is the only one taken into account. Looking to the REE grid code, the requirements upon a voltage disturbance at the PCC are basically two: Firstly, ensure that the generating facilities remain on-line (in order to a fast recovery to pre-fault values) and secondly help restoring system voltages.

This can be achieved by all the various wind turbine technologies, but exactly how it is achieved depends of the type of wind turbine technology. Without low-voltage ride-through

(LVRT) capability most wind turbine generators (irrespective of manufacturer or type) will trip during a system disturbance. So, the LVRT capability is a necessary feature on all wind turbine generators to remain the wind farms connected to the transmission grid.

For the case of wind turbines with full-scale converters, this kind of wind turbines supplies all the generated power through the converters. Thus, in the event of voltage dips in the grid, the grid side converter switches to reactive power priority, as a result, the active power injected by the converter can be reduced.

In short, if the wind turbine is able to ride through a disturbance and the necessary control and protection modifications have been made to ensure for fault ride-through, then the wind turbines will respond producing the required reactive power to restore voltage or at least a big part of it.

The considered scenario fits in this last case, all the wind turbines installed in the offshore wind farm can fulfill individually the REE grid code requirements for LVRT, as is verified in section 5.2.2.4. Consequently, an offshore wind farm composed by this kind of turbines with full-scale converters may be capable to fulfill the grid code requirements or at least be close to fulfilling such demands.

Therefore, in the present section, several simulations are carried out to analyze the behavior of the considered wind farm. The objective of this analysis based on simulations is to determine if the considered offshore wind farm (chapter 5) with this type of wind turbines can fulfill the grid code requirements and if it cannot, why not and how far is it. Then on the basis of the simulation results, purpose a solution to fulfill those requirements.

### 7.4.1 Evaluation of the grid code fulfillment of the considered wind farm for voltage dips at the PCC

In this section, firstly, the behavior of the developed wind farm model (see section 7.2) during a three-phase voltage dip at PCC is evaluated. With the objective to determine if it can fulfill the REE grid code requirements.

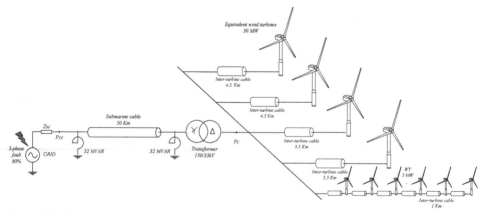

Figure 7.21 The simulation scenario to evaluate the behavior of the wind farm upon voltage dips.

For that purpose, upon the considered offshore wind farm, a three-phase voltage dip of 80% at the PCC (the most severe 3-phase fault considered by the REE grid code) is applied. To focus the analysis in the worst case is considered that all the wind turbines are operating at full-load (Table 5.13), at 90% of the rated power with a unity power factor. The simulated scenario is depicted in Figure 7.21. The voltage, current and power at the PCC obtained in the simulation of the considered scenario is shown at Figure 7.22.

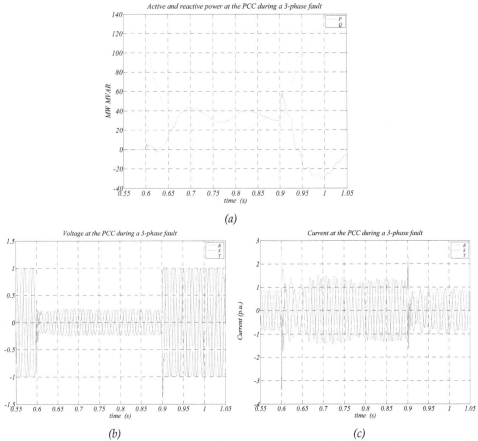

*(a)*

*(b)*                                                   *(c)*

Figure 7.22 Current, voltage and active /reactive power at the PCC for the considered simulation scenario upon 3-phase 80% dip.

In first place, the tripping of the wind turbines and the protection of its integrity due to the current / voltage peaks in the transmission system is discussed. Thus, the current peaks at the beginning of the fault and just in the clearance of the fault (Figure 7.22 (c)) are evaluated.

Those current peaks are exactly in the phases where the capacitive component of the submarine cable is charged and the magnitudes of the current peaks are also proportional to the phase voltages (see Figure 7.23).

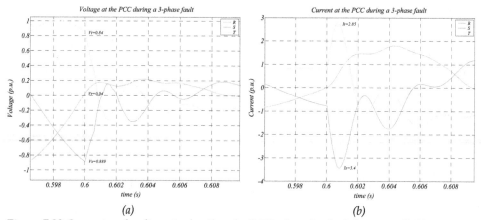

Figure 7.23 Current and voltage in detail at the PCC when the fault happens (0.6s).

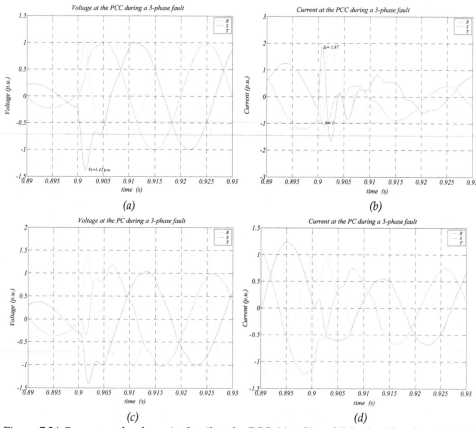

Figure 7.24 Current and voltage in detail at the PCC (a) – (b) and PC (c) – (d), when the fault is cleared (0.9s).

At the fault clearance occurs something similar, the bigger is the voltage step, the bigger is the needed current to charge the capacitive component of the cable. In other words, at the fault clearance, the phase voltage steps to the same point of the sinus but to the nominal voltage. Thus, the closer is the phase voltage to the maximum peak value, the bigger is the voltage step of this phase and the bigger is the inrush current, Figure 7.24 *(a) - (b)*.

Therefore, the circuit has the typical inrush currents for the energizing / de-energizing of the cable, in concordance with the results described in [126] and in section 7.1. So, it is possible to conclude that those current peaks are inherent to the capacitive behavior of the submarine cable upon fast voltage changes.

Looking to the voltage peaks (Figure 7.24, *(a) - (c)*), it is possible to see over voltages at the cable energizing, in the same way that is described in section 7.1 for the analysis of the submarine cable energizing.

The maximum value of the peaks and their duration is a key issue, because all the equipment of the wind farm must remain connected during the voltage fault. Therefore, in the next step of the evaluation, these voltage / current peaks are characterized

The voltage and current peaks in the main points of the offshore wind farm are summarized in Table 7.2 and the three-phase signals are displayed at Figure 7.22 and Figure 7.25. The duration of the voltage / current peaks are measured from the instant when the signal overcomes its nominal values to the instant when the signal comes back to its nominal values.

| | Current | | Voltage | |
|---|---|---|---|---|
| **Xsc=5%** | | | | |
| **PCC** | | | | |
| *Phase* | *Maximum peak* | *duration* | *Maximum peak* | *duration* |
| *Phase R* | *1.85 In* | *1.7 ms* | - | - |
| *Phase S* | *3.4 In* | *1.8 ms* | *1.4 Vn* | *1.5 ms* |
| *Phase T* | *2.8 In* | *1.4 ms* | - | - |
| **PC** | | | | |
| *Phase R* | - | - | *1.2 Vn* | *1.5 ms* |
| *Phase S* | - | - | *1.41 Vn* | *1.6 ms* |
| *Phase T* | - | - | *1.62 Vn* | *1.4 ms* |
| **At the WT terminals** | | | | |
| *Phase R* | - | - | *1.3 Vn* | *2 ms* |
| *Phase S* | - | - | *1.21 Vn* | *3 ms* |
| *Phase T* | - | - | *1.7 Vn* | *1.7 ms* |

Table 7.2 Characteristics of the voltage and current peaks at the main points of the simulation scenario.

The current amplitude oscillations during the maintenance of the fault, which can be seen at Figure 7.25 *(b)-(d)*, are caused by the voltage difference at the wind turbine terminals. Each wind turbine see a different impedance depending on its location in the inter turbine grid. In normal operation (rated voltage at the PC), this difference is not significant.

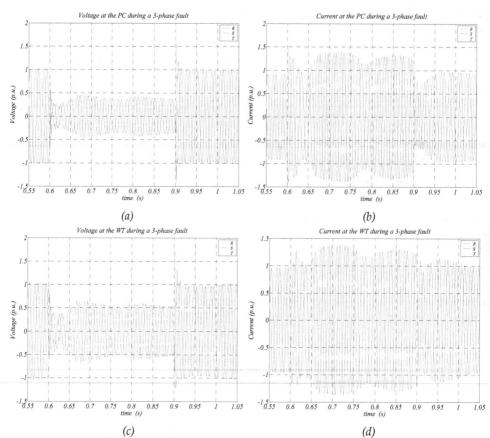

Figure 7.25 Current and voltage at the PC *(a) – (b)* and wind turbine terminals *(c) – (d)*, three-phase signals.

However, this difference increases in voltage dips, due to the fact that the wind turbines are supporting the voltage at the PC. Furthermore, this difference increases as a step when the fault occurs. Thus, the wind turbine converters need time to synchronize and stabilize the current at the PC.

The considered wind turbine model allows temporary over currents during the faults, consequently, the current during voltage dips increases to 1.2 p.u., as can be seen at Figure 7.25 *(b)-(d)*.

As is mentioned in section 7.1, the current/voltage peaks are provoked by the capacitive component of the submarine cable, these current peaks (submarine cable energizing / de-energizing inrush current peaks) depends on two variables: the voltage drop/step and the series impedance between the submarine cable and where the voltage drop/step occurs.

The voltage drop/step depends on the fault and the maximum depth is determined by grid codes, but the grid impedance depends on the point of the grid that the offshore wind farm is connected.

| Phase | Current | | Voltage | |
|---|---|---|---|---|
| | Maximum peak | duration | Maximum peak | duration |
| **Xsc=15%** | | | | |
| **PCC** | | | | |
| Phase R | 1.85 In | 0.7 ms | - | - |
| Phase S | 2.6 In | 3 ms | 1.36 Vn | 2 ms |
| Phase T | 1.77 In | 1.8 ms | - | - |
| **PC** | | | | |
| Phase R | - | - | 1.2 Vn | 2.8 ms |
| Phase S | - | - | 1.3 Vn | 2.3 ms |
| Phase T | - | - | 1.15 Vn | 1.2 ms |
| **At the WT terminals** | | | | |
| Phase R | - | - | 1.16 Vn | 3.3 ms |
| Phase S | - | - | 1.2 Vn | 2.6 ms |
| Phase T | - | - | 1.32 Vn | 1.8 ms |

Table 7.3 Characteristics of the voltage and current peaks at the main points of the simulation scenario.

In the present case, a strong grid point is chosen. Nevertheless, not all the wind farms have to connect to a strong grid. It is possible that some offshore wind farms have to connect for example to a point with bigger grid impedance.

In this way, to evaluate this point, the voltage and current peaks in the main points of the base offshore wind farm (Figure 7.21) considering a short circuit impedance of 15% is estimated. The results are summarized in Table 7.3

Regarding to Table 7.2 and Table 7.3, it can be seen how with bigger short circuit impedance, the maximum peak value in the fault transient decreases. So, the short circuit impedance has an important role on the cable energizing / de-energizing. In this way, using auxiliary devices to limit the current during fault conditions (such as current limiting reactors) [127], it is possible to achieve the same effect.

The current peaks appear only at the PCC with a short duration. Considering that the over current protection systems are adjusted for a specific thermal limit, the shorter is the peak, the less dangerous is for the system. Furthermore, the inrush currents of the transformer (as for the submarine cable) are 3-5 times bigger than the nominal current. Therefore, the protection devices have to be dimensioned for current peaks with these values.

On the other hand are the voltage peaks. Upon this type of peaks the most vulnerable equipments / components are those ones based on power electronics. Looking at the maximum peak value of the over voltages, 1.7 - 1.3 times the nominal voltage at the wind turbine terminals, these values are quite dangerous for the integrity of the power electronic devices. Thus, this kind of devices must be provided with the proper protection circuits.

However, considering that the voltage and current peaks measured in simulation are not dangerous for the main elements of the transmission system: Step-up transformer (not due to the peak value), submarine cable and inductive compensators.

Those transient over voltages are a matter of the wind turbine design (grid side converters protections), which is not considered as one of the objectives of the book. Consequently, the characterization of those protection devices is not analyzed.

### 7.4.1.1 Reactive power injection at the PCC during voltage dips

Apart from the protection of the integrity of the system and avoid the tripping of the wind turbines, the other main issue of the wind farms to fulfill the LVRT requirements is the injection of the proper reactive current at the PCC.

Looking at the results displayed at Figure 7.22, it is possible to conclude at first sight that the wind farm does not fulfill the grid code requirements. Due to the fact that the injected active and reactive-capacitive power to the distribution grid at the PCC is similar, unless predominately reactive-capacitive.

To analyze this aspect, the evolution of the active and reactive power in the main point of the offshore wind farm is depicted in Figure 7.26.

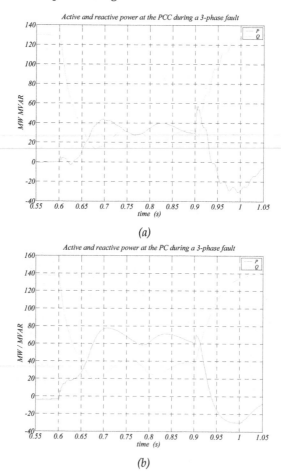

(a)

(b)

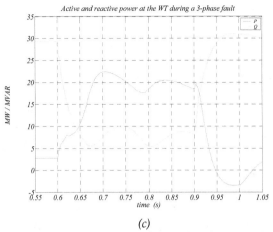

*(c)*

Figure 7.26 Active and reactive power at the main points of the simulation scenario, *(a)* PCC, *(b)* PC and *(c)* at the terminals of the wind turbine.

The offshore wind farm does not accomplish with the required reactive current injection for two reasons: The wind turbine does not inject the proper reactive current in relation to the active current (Figure 7.26 *(c)*) and the transmission system generates inductive reactive current (Figure 7.26 *(a)* - *(b)*).

The behavior of the wind turbines upon voltage dip is verified at 5.2.2.4. However, as can be seen in Figure 7.26 *(c)*, the wind turbines do not inject the reactive current in the correct quantities. One reason for this reduction is the difference between the voltage at the terminals of the wind turbines and the PCC. The voltage at the wind turbine terminals is significantly higher than at the PCC.

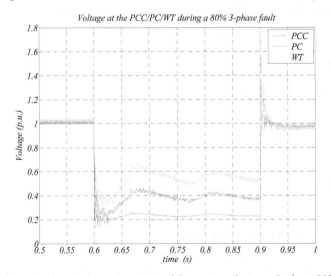

Figure 7.27 Voltage module at the main points of the system during a 3-phase 80% voltage dip.

The wind turbines have in their terminals a 50% voltage dip unless the 80% of the PCC, as can be seen in Figure 7.27.

Continuing with the analysis, why the transmission system generates inductive reactive power (Figure 7.26 (a) - (b)) is evaluated. For that purpose, the active and reactive power values of all the points of the transmission system are summarized on Table 7.4.

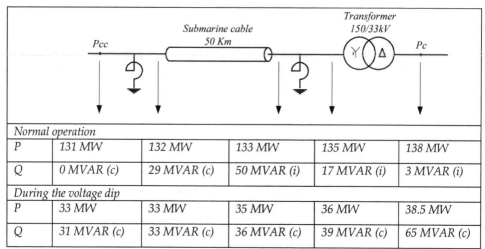

| Normal operation | | | | | |
|---|---|---|---|---|---|
| P | 131 MW | 132 MW | 133 MW | 135 MW | 138 MW |
| Q | 0 MVAR (c) | 29 MVAR (c) | 50 MVAR (i) | 17 MVAR (i) | 3 MVAR (i) |
| During the voltage dip | | | | | |
| P | 33 MW | 33 MW | 35 MW | 36 MW | 38.5 MW |
| Q | 31 MVAR (c) | 33 MVAR (c) | 36 MVAR (c) | 39 MVAR (c) | 65 MVAR (c) |

Table 7.4 Active and reactive power at all the points of the transmission system during an 80% 3-phase voltage dip, (i) inductive reactive power and (c) capacitive reactive power.

From the results displayed on Table 7.4, it is possible to see a reactive-capacitive power reduction between the two ends of the submarine cable during the voltage dip unless an increment. Therefore, with a huge voltage reduction, the submarine cable can generate inductive reactive power depending on the current through the cable.

Using the equations (98) and (99), see section 4.3.2.2, it is possible to estimate the reactive power generated in the submarine cable. Thus, considering a reduction of 80% on the applied voltage to the cable while flows the rated current, equations (145) and (146), the submarine cable generates 2.22 MVAR (0.74 MVAR per phase) of inductive reactive power (with rated current plus 10% 3 MVAR).

$$Q_C = \frac{|V_C|^2}{|X_C|} = \frac{|17.3kV|^2}{|273\Omega|} = 1MVAR \tag{145}$$

$$Q_L = |I_L|^2 \cdot |X_L| = |577\,A|^2 \cdot |5.24\Omega| = 1.74\,MVAR \tag{146}$$

## 7.5 Proposed electrical connection infrastructure

From the evaluation of the previous section is known that the voltage dips at the PCC does not causes dangerous over voltages for the offshore wind farms connection electric infrastructure (step-up transformer, cable, compensators, etc…), but these voltage peaks can be dangerous for the electronic devices.

The first step to fulfill the grid code requirements is to avoid the tripping of the system during grid faults. Thus, in the proposed solution is considered that the converters have well adjusted their protections. Consequently, it is considered that the offshore wind farm will not trip during voltage dips. However, the protection devices of the wind turbines are not considered in detail.

Therefore, this section is focused on the injection of the proper reactive-capacitive power at the PCC, the other main issue of the wind farm to fulfill the LVRT requirements, as is exposed in the previous section.

### 7.5.1 Modification of the Ireactive / Itotal curve of the wind turbines

As is shown in Figure 7.26, the wind turbines do not inject the required reactive current due to the voltage difference between the PCC and wind turbine terminals (Figure 7.27).

One solution for this problem can be the modification of the Ireactive / Itotal curve depending on the voltage of each wind turbine (see Appendix C: REE Grid Code Requirements for Voltage Dips Figure C.3). This modification has to be oriented to fulfill the grid code curve at the PCC, not at the wind turbine terminals. So, the voltage limits of the curve have to be adjusted considering the propagation of the voltage dip trough the transmission system.

To carry out the adjustment of the voltage limits for the Ireactive / Itotal curve on the considered scenario, several symmetric voltage dips with key depth are simulated in order to know how voltage dips are propagated. The simulation scenario to evaluate the voltage dips propagation is depicted on Figure 7.28.

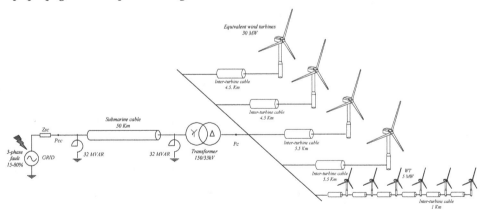

Figure 7.28 The simulation scenario to evaluate the behavior of the wind farm upon voltage dips.

The evolutions of the voltage through the electric infrastructure for voltage dips with the key depths are summarized in Table 7.5 and the Ireactive / Itotal ratio at the PCC for those voltage dips in Figure 7.29

| At PCC | At PC | At WT terminals |
|---|---|---|
| A 3-phase voltage dip with 80% of depth $Vpcc=0.2$ p.u. | 60 % of depth $Vpc=0.4$ p.u. | 48% of depth $Vwt=0.52$ p.u. |
| A 3-phase voltage dip with 50% of depth $Vpcc=0.5$ p.u. | 40% of depth $Vpc=0.62$ p.u. | 30% of depth $Vwt=0.73$ p.u. |
| A 3-phase voltage dip with 20% of depth $Vpcc=0.8$ p.u. | 17% of depth $Vpc=0.83$ p.u. | 13% of depth $Vwt=0.86$ p.u. |
| A 3-phase voltage dip with 15% of depth $Vpcc=0.85$ p.u. | 13% of depth $Vpc=0.87$ p.u. | 10% of depth $Vwt=0.90$ p.u. |

Table 7.5 Voltage module at different parts of the electric connection infrastructure for several voltage dips with key depths.

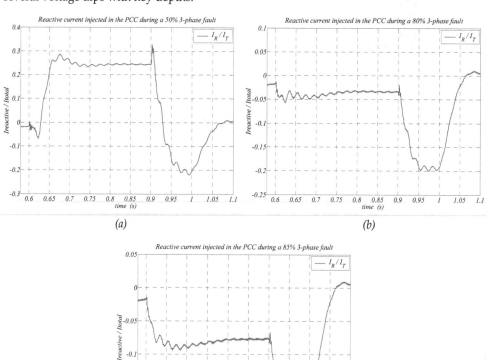

(a)                                          (b)

(c)

Figure 7.29 Ireactive / Itotal ratio at the PCC, (a) during a 3-phase 50% voltage dip, (b) during a 3-phase 20% voltage dip and (c) during a 3-phase 15% voltage dip.

Based on the simulation results illustrated in Table 7.5, for voltage dips with less than the 20% of depth, the wind turbine does not inject reactive power because the voltage at the wind turbine terminals is up to 0.85p.u. Consequently, it is necessary to change the voltage limit to start injecting reactive current to fulfill the REE requirements.

Looking at the results shown on Table 7.5, for voltage dips with 15% of depth at the PCC, there is a 10% voltage dip at wind turbine terminals. So, to fulfill the REE grid code, the wind turbines have to start injecting reactive power with only a 10% voltage drop at their terminals.

As regards to the last voltage limit modification, the voltage limit to start injecting the maximum reactive current, based on Table 7.5, is elevated to 0.75p.u. In this way, the modified Ireactive / Itotal curve to fulfill the REE requirements at the PCC using wind turbines is depicted in Figure 7.30.

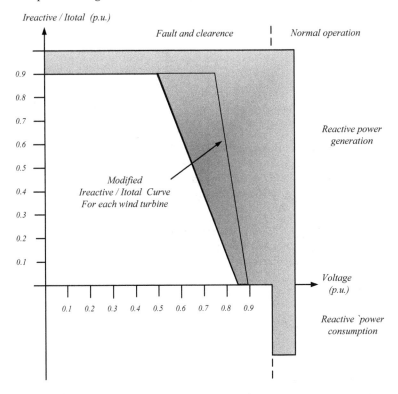

Figure 7.30 Modified Ireactive / Itotal curve depending on the voltage to fulfill the REE requirements at the PCC.

Even with this modification, the offshore wind farm does not fulfill the grid code requirements at the PCC, due to the fact that the transmission system generates inductive reactive power, see Table 7.4.

To verify this point, the same scenario of the section 7.4 with the same conditions (generating the 90% of the rated power) is simulated, but this time, the wind turbines have modified their Ireactive /Itotal curve (see Figure 7.30). The simulation results of the scenario shown in Figure 7.21 with modified Ireactive /Itotal curve are depicted at Figure 7.31.

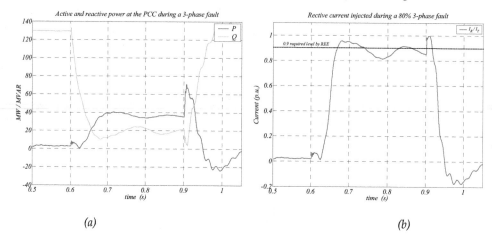

(a)                                                    (b)

Figure 7.31 Active and reactive power at the PCC (a) and Ireactive / Itotal at the PCC (b) for the considered offshore wind farm with modified Ireactive / Itotal curve.

As can be seen from the results displayed on Figure 7.31, even with the modified Ireactive / Itotal curve, the wind farm does not fulfill the REE grid code requirements. The reason for not fulfill code requirements is the inductive reactive power generated in the transmission system. At the PC the wind farm injects the proper quantity of capacitive reactive power, but at the PCC does not (Table 7.6).

To solve this problem, one possibility is to increase the size of the DC chopper. In this way, the DC chopper can be able to consume more active power and the wind turbines will inject less active power and more capacitive reactive power.

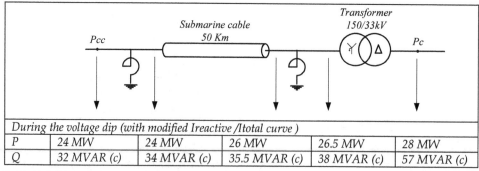

| During the voltage dip (with modified Ireactive /Itotal curve ) | | | | |
|---|---|---|---|---|
| P | 24 MW | 24 MW | 26 MW | 26.5 MW | 28 MW |
| Q | 32 MVAR (c) | 34 MVAR (c) | 35.5 MVAR (c) | 38 MVAR (c) | 57 MVAR (c) |

Table 7.6 Active and reactive power at all the points of the transmission system during a 80% 3-phase voltage dip (With modified Ireactive / Itotal curve), (i) inductive reactive power and (c) capacitive reactive power.

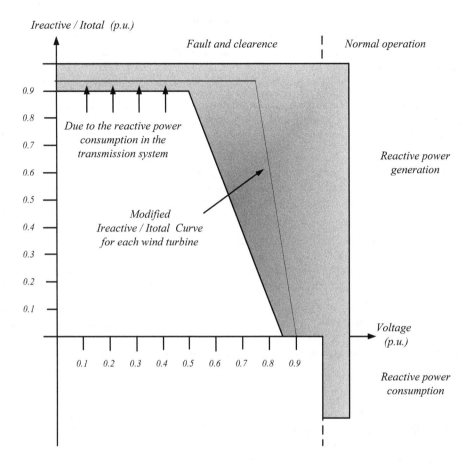

Figure 7.32 Modified Ireactive / Itotal curve depending on the voltage to fulfill the REE requirements at the PCC by reducing the injected active power during faults.

### 7.5.2 Characterization of auxiliary equipment, STATCOM

Another option is the use of a STATCOM [128] at the PCC as an auxiliary system. The wind turbines of the offshore wind farm can inject the main amount of the reactive power and reduce the generated active power during a voltage dip at the PCC (Figure 7.30), then, at the PCC the auxiliary equipment can inject the rest of the capacitive reactive power. This STATCOM also can provide some other services like: power factor control.

Moreover, using a STATCOM at the PCC, it can provide the required capacitive reactive power for voltage dips with less than the 20% of depth. This means that the voltage limit of the wind turbines to start injecting reactive power can remain at V=0.85 p.u. without increments (Figure 7.33). As a result, for any voltage reduction at the inter turbine grid, for example of the 10 % (not caused by a voltage reduction at the PCC), the wind turbines will not prioritize the injection of reactive current.

Focusing on the STATCOM, if there is considered that Ireactive / Itotal curve is only modified in voltage (only voltage limits adjusted, Figure 7.30), the STATCOM has to compensate the inductive-reactive power generated at the transmission system during voltage dips. This means that the rated power of the STATCOM depends on the inductive component of the submarine cable and the leakage inductance of the step-up transformer.

Therefore, the less are this inductive components, the less is the generated inductive reactive power at the transmission system and the less is the necessary rated power of the STATCOM to fulfill the grid codes.

An option to reduce the needed reactive power at the PCC during voltage dips is to provide thyristors to the onshore static compensation, giving them the ability to disconnect themselves when a fault occurs. The other way is the reduction as far as possible the leakage inductance of the step-up transformer at the offshore platform.

In short, the size of the STATCOM depends on the inductive reactive energy generated in the transmission system during the voltage dip and the amount of the active / reactive power injected by wind turbines, i.e. the modification of the Ireactive / Iactive curve of the wind turbines by increasing their injected reactive current. Therefore, many options to inject the proper reactive power at the PCC are allowed.

However, due to the huge amount of inductive reactive power generated in the transmission system, 25 MVAR in the worst case (80% symmetric voltage dip, 20% of residuary voltage), Table 7.4. The use of a STATCOM to compensate the reactive power generated in the transmission system will require a 125 MVAR rated power.

Consequently, the proposed solution priorities the use of the grid side converters of the wind turbines to reduce active power and inject the biggest part of the required reactive power.

Considering the voltage dip propagation through the transmission system of the considered scenario, for voltage dips with less depth than 20%, the voltage at the wind turbine terminals is up to 0.85p.u. Therefore, to keep the voltage limit of the wind turbines to start injecting reactive power at 0.85 p.u., the STATCOM has to provide the required reactive power for voltage dips with less depth than 20%

The Ireactive / Itotal requirement at the PCC for a voltage dip with 20% of depth (Appendix C: REE Grid Code Requirements for Voltage Dips) is 0.128. Consequently, for the worst case, when the wind farm is generating the rated active power (injecting the rated active current) the required reactive current is 0.13 p.u., equations (147) - (149).

$$I_{reactive} / I_{total} = \frac{I_{reactive}}{\sqrt{I_{reactive}^2 + I_{active}^2}} \tag{147}$$

$$I_{reactive} / I_{total} = 0.128 = \frac{I_{reactive}}{\sqrt{I_{reactive}^2 + 1^2}} \tag{148}$$

$$I_{reactive} = \sqrt{\frac{0.128^2}{1 - 0.128^2}} = 0.13 \qquad (149)$$

In this way, by using a STATCOM of 20 MVAR (0.13 p.u.) rated power at the PCC, the change of the minimum voltage limit of the wind turbines to prioritize the reactive power injection is not necessary.

The model and characteristics of the considered STATCOM used to verify the proposed electric connection infrastructure are summarized in Appendix G: Considered STATCOM Model to Validate the Proposed Solution.

### 7.5.3 Proposed solution

For the proposed solution, a 20 MVAR STATCOM (13%) and the modification of the Ireactive / Itotal curve depending on the voltage shown at Figure 7.33 is considered.

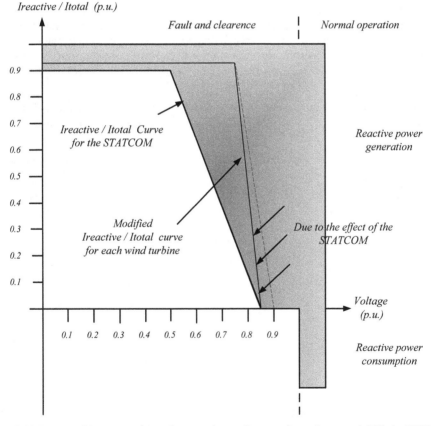

Figure 7.33 Proposed Ireactive / Itotal curve depending on the voltage to fulfill the REE requirements at the PCC.

Therefore, in the proposed solution, the compensation of the inductive reactive power generated in the transmission system is made by grid side converters of the wind turbines. Consequently, the performed Ireactive / Itotal curve modification, the increment in the Ireactive / Itotal relation depends on the inductive reactive power generated in the transmission system.

Thus, in order to keep the Ireactive / Itotal relation as low as possible, there is considered that the onshore compensation can be disconnected during voltage dips at the PCC, reducing in 2 MVARs the required reactive power at PCC. As a result, for the new curve the maximum Ireactive / Itotal relation is modified from 0.9 to 0.93.

Nevertheless, for the purposed complete scheme, besides the STATCOM and onshore variable inductive compensations, all the proposed solutions for the problems discussed in previous sections are taken into account: The fixed inductive reactive compensators at both ends of the submarine cable adjusted for the case when the wind farm is transmitting the rated power (32 MVAR + 32 MVAR for 50 Km of the selected cable) and the passive filters for the main resonance of the system (330 Hz).

The transient over voltages and over currents do not affect to the transmission system elements such as the step-up transformer or the submarine cable. Nevertheless, the more delicate devices (power electronic converters) need to be protected. Thus, for the considered scenario is considered that the wind turbines and the STATCOM are provided with those devices.

Therefore, depending on those protection devices, the passive filters evaluated in section 7.3.3.1 to improve the transient response of the system may be not required. However, these passive filters are considered for the proposed solution.

In short, the complete electric scheme of the purposed AC offshore wind farm is depicted in Figure 7.34.

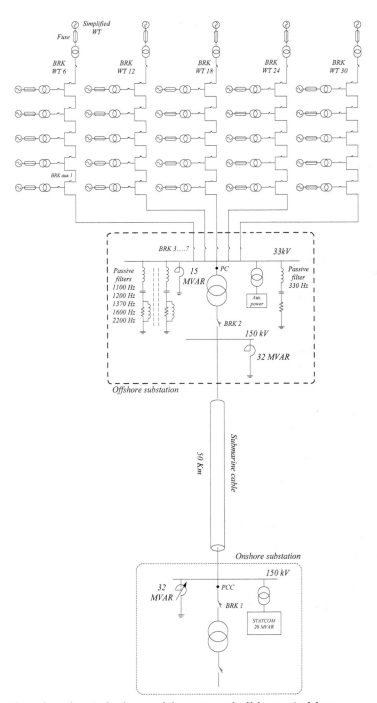

Figure 7.34 Complete electrical scheme of the proposed offshore wind farm.

## 7.5.4 Verification via PVVC of the purposed wind farm

For testing and validation of wind turbines, REE has defined a procedure detailing all the tests and characteristic in the validation process, the PVVC (Procedimiento de verificación, validación y certificación de los requisitos del PO 12.3 sobre la respuesta de las instalaciones eólicas ante huecos de tensión) [101]. Therefore, as is made in section 5.2.2.4 for wind turbines, in the present section, this procedure with the same conditions is applied to the entire offshore wind farm.

Thus, upon the proposed wind farm, the four faults defined in the PVVC are applied (Table 5.12). Then, the voltage and current is measure at the PCC to analyze if the proposed wind farm accomplishes with the REE requirements.

Results of a fault category 1: Three-phase fault - partial load

| | Limit P.O. 12.3 | Test results |
|---|---|---|
| *Net reactive power consumption, in cycles of 20ms, during a period of 150ms after the beginning of the fault:* | -0.1500 | 0 |
| *Net reactive power consumption, during a period of 150ms after the clearance of the fault:* | -0.0900 | 0 |
| *Net reactive current consumption, in cycles of 20ms, during a period of 150ms after the clearance of the fault:* | -1.5000 | 0 |
| *Net active power consumption during the fault:* | -0.1000 | 0 |
| *Net reactive power consumption during the fault:* | -0.0500 | 0 |
| *Fulfillment of the $I_{reactive}/I_{total}$ requirement:* | 0.9000 | 0.9367 |

Table 7.7 Summarized results of a 1st category fault for the test defined in the PVVC.

Results of a fault category 2: Three-phase fault - full load.

| | Limit P.O. 12.3 | Test results |
|---|---|---|
| *Net reactive power consumption, in cycles of 20ms, during a period of 150ms after the beginning of the fault:* | -0.1500 | 0 |
| *Net reactive power consumption, during a period of 150ms after the clearance of the fault:* | -0.0900 | 0 |
| *Net reactive current consumption, in cycles of 20ms, during a period of 150ms after the clearance of the fault:* | -1.5000 | 0 |
| *Net active power consumption during the fault:* | -0.1000 | 0 |
| *Net reactive power consumption during the fault:* | -0.0500 | 0 |
| *Fulfillment of the $I_{reactive}/I_{total}$ requirement:* | 0.9000 | 0.9164 |

Table 7.8 Summarized results of a 2nd category fault for the test defined in the PVVC.

Graphic results:

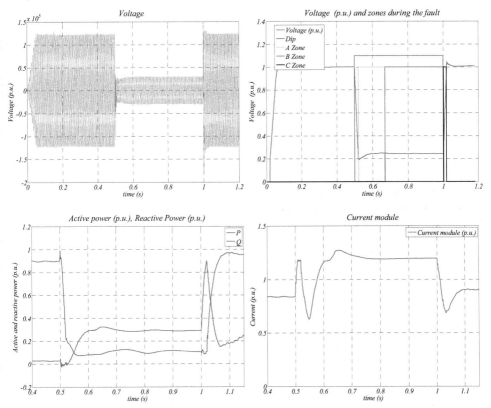

Figure 7.35 Summarized graphical results of a 2nd category fault for the test defined in the PVVC, voltage (module and signals), power and current results.

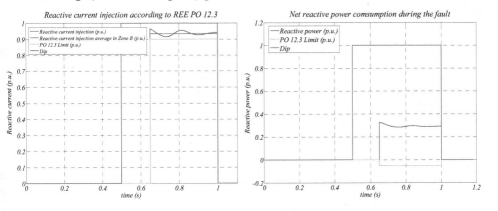

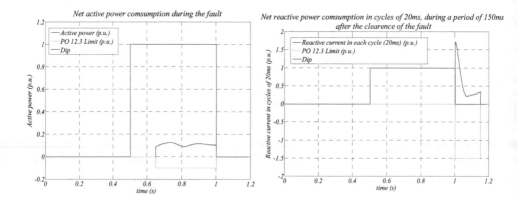

Figure 7.36 Summarized graphical results of a 2nd category fault for the test defined in the PVVC, reactive current and power consumption in B zone and reactive current consumption in C zone.

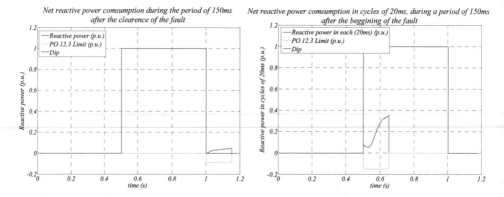

Figure 7.37 Summarized graphical results of a 2nd category fault for the test defined in the PVVC, reactive power consumption in C zone and A zone.

Results of a fault category 3: Two phase ungrounded fault - partial load

|  | Limit P.O. 12.3 | Test results |
|---|---|---|
| Net reactive power consumption, in cycles of 20ms, during the maintenance of the fault: | -0.4000 | 0 |
| Net reactive power consumption, during the maintenance of the fault: | -0.0400 | 0 |
| Net active power consumption, in cycles of 20ms, during the maintenance of the fault: | -0.3000 | 0 |
| Net active power consumption, during the maintenance of the fault: | -0.0450 | 0 |

Table 7.8. Summarized results of a 3rd category fault for the test defined in the PVVC.

Results of a fault category 4: Two phase ungrounded fault - full load

| | Limit P.O. 12.3 | Test results |
|---|---|---|
| Net reactive power consumption, in cycles of 20ms, during the maintenance of the fault: | -0.4000 | 0 |
| Net reactive power consumption, during the maintenance of the fault: | -0.0400 | 0 |
| Net active power consumption, in cycles of 20ms, during the maintenance of the fault: | -0.3000 | 0 |
| Net active power consumption, during the maintenance of the fault: | -0.0450 | 0 |

Table 7.9 Summarized results of a 4th category fault for the test defined in the PVVC.

Graphic results:

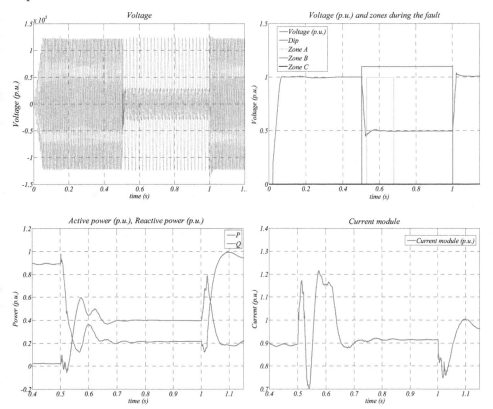

Figure 7.38 Summarized graphical results of a 4th category fault for the test defined in the PVVC, voltage (module and signals), power and current results..

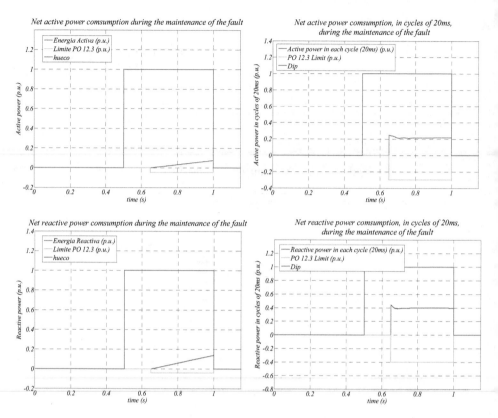

Figure 7.39 Summarized graphical results of a 4th category fault for the test defined in the PVVC, active and reactive power consumption in B zone.

As can be seen from Figure 7.35 - Figure 7.39, the proposed wind farm fulfills the reactive power and the fast recovery (to the pre-fault values) requirements of the REE grid code. Therefore, the proposed offshore wind farm is suitable to connect to a distribution grid operated by REE.

## 7.6 Chapter conclusions

The electric connection infrastructure of the offshore wind farm is composed by several inductive and capacitive elements. Due to this fact, the energizing of the circuit and the transient responses are especially of concern.

The analysis performed in the present chapter reveals that a fault clearance (by the circuit breaker) in a feeder of the inter-turbine grid can damage the power electronic devices connected to this grid. The short circuit current interruption causes an over-voltage of 50% in the collector point.

Besides, the faults or voltage dips at the PCC which provokes the de-energizing and energizing of the submarine cable also causes transient over-voltages. Therefore, all the power electronic devices connected to the electric connection infrastructure of the offshore wind farm must be provided with the proper protective devices (surge arresters, etc…) to ride through those transient over-voltages.

In the same way, some studies give description of the phenomenon that produces high over voltages internally in the transformer winding caused by high frequency transients [121], which apparently are responsible for the transformer insulation failures. Thus, also the transformers have to be provided with the proper protective devices.

The use of passive filters to minimize the voltage peaks and the high frequency components of the transients can be an option. But, as is mentioned before, the passive filters definition is based on the perfect knowledge of parameters. Therefore, in real conditions, where the parameters are not well known, passive solutions might not be suitable.

Another point evaluated in this chapter is the behavior of the offshore wind farm upon voltage dips at the PCC, from the point of view of the REE grid code requirements. To connect an offshore wind farm to the distribution grid has to fulfill the rules established by the system operator. One important requirement is the reactive power injection to the main grid during voltage dips at the PCC.

To achieve this requirement is important to know the behavior of the submarine cable, because this element can generate inductive reactive power or capacitive reactive power depending on the applied voltage and the current through the cable.

Due to this fact, depending on the submarine cable characteristics, the cable and its reactive power compensators (see chapter 4, section 4.3) both of them can generate inductive reactive power during a voltage dip, making harder to fulfill the grid code requirements.

Nevertheless, full-converter wind turbines and their capability to inject reactive current can help to fulfill the reactive current requirement at the PCC generating the biggest part of the required reactive current.

As a result, using full-converter wind turbines and an STATCOM as auxiliary equipment at the PCC is possible to fulfill the REE grid code requirements, as can be conclude from the results of the section 7.5.4.

# Chapter 8

# Conclusions and Future Work

From the analysis carried out in this report is clear that it is possible the energy transmission from the offshore wind farms via AC transmission within some rated power and rated voltage boundaries.

To focus the analysis in a specific case, there is analyzed the transmission of the 150 MW at 150kV to 50Km to the shore. In this way, upon this configuration several key issues of the AC transmission systems are analyzed.

Firstly, there are analyzed submarine cables, the way to represent electrically submarine cables and several options to model the cable.

Once the submarine cable has been modeled and validated. The first key issue to analyze is the management of the reactive power flowing through the submarine cable. This analysis shows that a proper reactive power management can reduces significantly the active power losses and the voltage drop in the submarine cable. Furthermore, different ways to achieve this reactive power management are compared, such as the use of fixed inductances at both ends of the line.

So, comparing the reactive power management based on fixed inductances at both ends of the line, with the variable compensation at both ends. It is possible to see that the both methods have similar improvements. For both ways, the maximum voltage drop, the maximum active power losses and the required maximum current for the cable are exactly the same. As a result is achieved as a conclusion that the fixed inductances at both ends are the best option to manage the reactive power through the submarine cable.

Another key issue of the AC transmission configurations is the undesired harmonic amplifications due to the resonances of the system. Thus, using the developed offshore wind farm model as the base scenario, the analysis of its frequency response is carried out. Estimating the resonances of the system and measuring the influence of the main components of the transmission system in its frequency response.

From the results of this study, it is possible to observe that the offshore wind farms have the potential to amplify low order harmonics due to the iteration between the capacitive component of the submarine cable and the leakage inductance of the step-up transformer.

At this point, highlight that the present analysis as a first approximation does not takes into account the effect of the control strategies of the wind turbines. There are not considered control strategies oriented to mitigate filters avoiding the resonance, and neither is considered further and more complex analysis about the harmonic risk or problems related to the iteration between the control strategies of the wind turbines

As any other installation, offshore wind farms must be protected against different eventualities. Moreover, the behavior of these types of generation plants during any eventuality, which causes a voltage drop at the point of common coupling (PCC), is crucial to ensure the quality and the continuity of the electric supply. Therefore, if this kind of generation systems wants to connect to the distribution grid, they have to satisfy the grid code requirements of the system operator (SO).

In conclusion, it is not enough the normal operation analysis to carry out the pre-design of an offshore wind farms electric system. Therefore, to complete the analysis of the AC transmission configurations, an analysis about the disturbances in the electric connection infrastructure and their effects is carried out.

The analysis performed reveals that the switching actions in the electric connection infrastructure have the potential to cause over current and over voltages. The cause of those transient over currents and voltages is the composition of the electric connection infrastructure itself.

In addition to inductive elements, the electric connection infrastructure of the offshore wind farm has huge capacitive elements, such as the transmission submarine cable or the inter-turbine cable.

In the same way, faults or dips at the PCC, which provokes the de-energizing and energizing of the submarine cable also causes dangerous transient over-voltages. Due to this fact, the power electronic devices of the offshore wind farm must be provided with the proper over-voltages protection system.

Another point evaluated in this book is the behavior of the offshore wind farm upon voltage dips at the PCC, from the point of view of the REE grid code requirements.

As one of the most important requirement, the reactive power injection to the main grid during voltage dips in the PCC is analyzed. The submarine cable generates a huge amount of capacitive reactive power in normal operation. Nevertheless, if the voltage applied to the cable drops, the submarine cable can generate inductive reactive power.

As a result, the submarine cable does not help injecting capacitive reactive power during voltage dips at the PCC. Moreover, as the cable can generate inductive reactive power, the offshore wind farm or its auxiliary equipment have to be dimensioned to inject at the PCC the required capacitive reactive power plus compensate the inductive reactive power generated at the cable.

In this way, the huge amount of inductive reactive power generated in the transmission system makes unattractive the use of a STATCOM to inject the main amount of reactive power during voltage dips.

However, full-converter wind turbines can be very helpful to fulfill the grid code requirements at the PCC. But, due to the fact that the wind turbines are located far away from the PCC, their reactive current injection characteristics for voltage dips have to be adjusted.

Therefore, the proposed solution priorities the use of the grid side converters of the wind turbines to reduce active power and inject the biggest part of the required reactive power and the STATCOM only as auxiliary equipment.

Note that the objective of the present book is the development of a methodology or a sequence of analyses, to make the first approach to the design of the offshore wind farms electric connection infrastructure. In other words, the objective is the identification of the problematic aspects of the energy transmission and grid integration by applying the sequence of analyses.

As a result, the present book does not define in detail several advanced design features or technological aspects. Thus, a more deep analysis of the identified problematic aspects is needed to design an advanced or detailed electric connection infrastructure.

In this way, from the analyses carried out, it is seen that the protection system of the converters has to be evaluated more carefully. Due to the fact that the breakers operations or voltage dips at the PCC can cause dangerous voltage peaks for them in the offshore wind farms transmission system.

Finally, this research is based on a specific case which is intended to be representative. So, the proposed solution has been dimensioned to fulfill REE grid code requirements. Consequently, as there are many other system operators and grid codes, the adaptation of the proposed solution to other grid codes will give consistency to this solution. In the same way, further analyses for other specific cases (other rated powers and lengths to shore) will complete this evaluation.

# References

[1] "Europe's onshore and offshore wind energy potential - An assessment of environmental and economic constraints," European environmental agency, 2009.

[2] "Wind in power, 2010 european statistics," EWEA, 2011.

[3] "Oceans of opportunity, Harnessing Europe's largest domestic energy resource," EWEA, 2009.

[4] "Pure power, Wind Energy Scenarios up to 2030," European Wind Energy Association, 2008.

[5] "Offshore Wind Energy: Action needed to deliver on the Energy Policy Objectives for 2020 and beyond," Commission of European communities, Brussels, 2008.

[6] European Commision.
http://ec.europa.eu/energy/electricity/package_2007/index_en.htm

[7] European Commision. http://ec.europa.eu/energy/climate_actions/index_en.htm

[8] Bureau de coordination energies renouvelables. http://www.wind-eole.com/fileadmin/user_upload/Downloads/Offshore/European_Offshore_Wind_Map_2009.pdf

[9] S. Chondrogiannis, M. Barnes, "Technologies for integrating wind farms to the grid (Intering report)", DTI, 2006.

[10] R. Ianini, J. Gonzalez, S. Mastrangelo, "Energía eólica, teoría y características de instalaciones," Boletín energético N°13, 2005.

[11] G.Jimenez, "Identificación de sitios eólicos," IV foro regional: Energía eólica y otras renovables en el futuro, 2004

[12] J. Moragues, A. rapallini, "Energía eólica," Instituto argentino de la energía, 2003

[13] "European wind atlas," Risø National Laboratory, 1989

[14] The Danish Wind Industry Association (DWIA). http://guidedtour.windpower.org

[15] J. Manuel Ruiz, "Energía eólica," Seminario de energías renovables, 2005.

[16] A. Perez, "Energía eólica," Universidad de Valladolid

[17] J.M. Escudero, "Manual de energía eólica," Ediciones mundi-prensa, 2008

[18] "Large scale integration of wind energy: analisys, issues and recommendations" EWEA 2005

[19] N. Jenkins, "Network integration and modeling of large wind turbines," CIRED, 2007

[20] A. Herrera, "Energía eólica marina," Jornadas de energía renovable marina, 2009

[21] "Energía Eólica," IDEA 2006

[22] Windatlas. http://www.windatlas.dk/Europe/About.html

[23] "The European offshore wind industry key trends and statistics 2009," EWEA, 2010

[24] F. Zimmermann, K. Biebler , "UK Offshore Wind Supply Chain Seminar," Copenhagen, September, 2009

[25] L.H. Kocewiak, C.L. Bak, J. Hjerrild, "Harmonic aspects of offshore wind farms," Seminar on Detailed Modelling and Validation of Electrical Components and Systems Denmark, February 2010

[26] Å. Larsson, A. Petersson, N. Ullah, Ola Carlson, "Krieger's Flak Wind Farm", Nordic wind power conference, May 2006

[27] W. Grainger, N. Jenkins, "Offshore Wind farm Electrical Conection Options," 1998

[28] S.D. Wright, A.L. Rogers, J.F. Manwell, A. Ellis, "Transmision options for offshore wind farms in the united states," AWEA 2002

[29] "Wind turbine Grid connection and Interaction," Deutsches Windenergie-Institut, 2001

[30] L. Hojbjerg Nielsen, "Grid development and integration", Energinet. 2007

[31] T. Ackermann, "Transmission System for Off-shore Wind Farms", KTH.

[32] S. Bozhko, R. Blasco, R. Li, J.C. Clare, G.M. Asher, "Control of Off-shore DFIG-Based Wind Farm Grid With Line-Commutated HVDC Connection," *IEEE Trans. On Energy Conv.* 2007.

[33] S. Lundberg, "Evaluation of wind farm layouts," *EPE Journal (European Power Electronics and Drives Journal)*, vol. 16, pp. 14-20, 2006.

[34] J. Green, A. Bowen, L.J. Fingersh, and Y. Wan, "Electrical Collection and Transmission Systems for Offshore Wind Power," NREL 2007.

[35] N. Barberis Negra, J. Todorovic, T. Ackermann, "Loss evaluation of HVAC and HVDC transmission solutions for large offshore wind farms" *Science direct* Electric Power Systems Research 76 (2006) 916–927

[36] S. Meier, "Novel Voltage Source Converter Based HVDC Transmission System for Offshore Wind Farms," Royal Institute of Technology, 2005.

[37] "Offshore wind energy grid connections," ABB 2003.

[38] P. D. Hopewell, F. Castro-Sayas, and D. I. Bailey, "Optimising the Design of Offshore Wind Farm Collection Networks," in *Universities Power Engineering Conference, 2006. UPEC '06. Proceedings of the 41st International*, 2006, pp. 84-88.

[39] S. Alepuz, S. Busquets-Monge, J. Bordonau, J.A. Martinez-Velasco, C.A. Silva, J. Pontt and J. Rodriguez, "Control strategies based on symmetrical components for grid connected converters under voltage dips," *IEEE Transactions on Industrial Electronics*, Vol: 56 Issue: 6, pp. 2162 – 2173, 2009

[40] L.Nian, "Transients in the Collection Grid of a novel Wind Farm Topology," Msc Thesis, KTH, Sweden, April, 2009

[41] P. Gadner, L.M. Craig, G.J. Smith, "Electrical Systems for offshore wind farms," Garrad Hassan & partners, UK 2006

[42] W.A. Thue, "Electrical power cable engineering," Marccel Dekker, 2003

[43] B.M. Weedy, "Sistemas electronicos de gran potencia," Editorial Reverte, 1981

[44] J.L. Jimenez , N.A. quino, I. Campos, "Entendiendo la autoinductancia," *Contactos 64*, pp. 59-6 3, 2007

[45] ABB, "XLPE cable systems, user's guide," rev 2

[46] L.C. Cisneros, "Cálculo de parámetros y operación en estado estacionario de líneas de transmisión," Modelado y operación de líneas de transmisión, ITM.

[47] B.M. Weedy, B.J. Cory, "Electric power systems," Wiley, 1998

[48] Y.L. Jiang, "matematical modelling on RLCG transmission lines," Xi'an Jiantong University, 2005

[49] M.C.M. Sánchez, "Medida de párametros de ruido de dispositivos activos, basadoa en fuente adaptada," Ph. D Thesis, UPC, 2003

[50] L.H-Restrepo, G.Caicedo delgado, F. Castro-Aranda, "Modelos de línea de transmisión para transitorios electromagnéticos en sistemas de potencia," Revista *Energía y computación* Vol 16 No 1 2008 p.21-32

[51] "Transients in power Systems," ECE524, Seasson 18, 2008

[52] L. H. Restrepo," Modelos de línea de transmisión para transitorios electromagnéticos en sistemas de potencia," Trabajo de Grado en Ingeniería Eléctrica, Universidad del Valle, Cali-Colombia, 2008.

[53] "Guidelines for representation of network elements when calculating transients," CIGRE Working Group, 1990.

[54] M. Khatir, S.Zidi, S.Hadjeri, M.K. Fellah, "Comparison of HVDC line models in PSB/SIMULINK based on steady-state and transients considerations," Acta Electrotechnica et Informatica Vol. 8, No 2, 2008, 50-55.

[55] "The Bergeron Method, A Graphic Method for Determining Line Reflections in Transient Phenomena," Texas instruments, 1996.

[56] M.H.J. Bollen, G.A.P Jacobs, "Implementation of an algorithm for traveling-wave-based directionat detection," Eindhoven university of technology, 1989.

[57] A. M.Elhaffar, "Power transmission line fault location based on current traveling waves," TKK Dissertations, 2008

[58] J. Arrillaga, L.I. Eguíluz, "Armónicos en sistemas de potencia," servicio de publicaciones de la universidad de Cantabria, 1994

[59] J. Marti, "Accurate Modeling of frequency dependent transmission lines in electromagnetic transients simulation," IEEE *transactions on power apparatus and systems*, 1982

[60] C. Dufour, W. Le-Huy," Highly Accurate modeling of frequency-dependent balanced transmission lines," *IEEE Trans. on Power Delivery*, vol. 15, No. 2, pp. 610-615,April, 2000

[61] F. Castellanos, J. R. Marti, F. Marcano, "Phase-domain multiphase transmission line models," *International Journal of Electrical Power & Energy Systems, Elsevier Science Ltd.* vol. 19, No. 4, pp. 241-248,May, 1997.

[62] F. Marcano," Modeling of transmission lines using idempotent decomposition," MASc. Thesis, Department of Electrical Engineering, The University of British Columbia, Vancouver, Canada, August, 1996

[63] B. Gustavsen, G. Irwin, R. Mangelrod, D.Brandt, K. Kent," transmission line models for the simulation of interaction phenomena between parallel AC and DC overhead lines," IPST 99 Procedings, pp. 61-67, 1999

[64] PSCAD, "User's guide," 2003.

[65] L. Nian, "Transients in the Collection Grid of a novel Wind Farm Topology," KTH Master of Science Thesis, 2009

[66] S. Meier, "System Aspects and Modulation Strategies of an HVDC-based Converter System for Wind Farms," Ph. D. thesis, KTH Stockholm, 2009, ISBN 978–91–7415–292–0.

[67] S. Lundberg, "Wind farm configuration and energy efficiency studies series DC versus AC layouts," Thesis, Chalmers University of Technology 2006.

[68] L. P. Lazaridis, "Economic comparison of HVAC and HVDC solutions for large offshore wind farms under special consideration of reliability," Msc. Thesis, KTH, 2005.

[69] S.D. Wright, A.L. Rogers, J.F. Manwell, A. Ellis, "Transmision options for offshore wind farms in the united states," AWEA 2002

[70] P. Gardner, L.M. Craig and G.J. Smith, "Electrical Systems for offshore wind farms," BWEA wind energy, 1998.

[71] P. Djapic, G. Strbac, "Further análisis for design of offshore infraestructure," Centre for Sustainable Electricity and distributed generation 2008

[72] S. Lundberg, "Evaluation of wind farm layouts," *EPE Journal (European Power Electronics and Drives Journal)*, vol. 16, pp. 14-20, 2006.

[73] R. Svoma, M.Dickinson, C. Brown, "Subsea connections to high capacity offshore windfarms: Issues to consider," CIRED 2007

[74] "BWEA Briefing Sheet: Offshore Wind," British wind energy association 2005

[75] BVG Associates ,"UK Offshore Wind: Moving up a gear," British wind energy association 2007

[76] "Gamesa G128-4.5MW," Gamesa, February 2010

[77] Vestas. http://www.vestas.com/en/wind-power-plants/procurement/turbine-overview/v112-3.0-mw-offshore.aspx#/vestas-univers

[78] Repower. http://www.windpowerengineering.com/tag/repower/

[79] M. Robinson, W. Musial, "Offshore wind technology overview," NREL/PR-500-40462, 2006

[80] P. Maibach, A. Faulstich, M. Eichler and S. Dewar, "Full-Scale Medium-Voltage Converters for Wind Power Generators up to 7 MVA," ABB, 2007

[81] "Ingedrive MV 300, Frequency converter data sheet," Ingeteam 2011

[82] A. Teninge, D. Roye, S. Bacha, J. Duval, "Low voltage ride-throufh capabilities of wind plant conbining different turbine technologies," EWEC, 2009.

[83] R. Mittal, K. S. Sandhu and D. K. Jain, "Low voltage ride through (LVRT) of grid interfaced wind driven PMSG," *Asian Research Publishing Network*, vol. 4, No. 5, July 2009

[84] S. Alepuz, S. Busquets-Monge, J. Bordonau, J. Gago, D. Gonzalez and J. Balcells, Interfacing Renewable Energy Sources to the Utility Grid Using a Three-Level Inverter," *IEEE Transactions on Industrial Electronics*, v 53, n 5, 1504-11, Oct. 2006

[85] R.C. Portillo, M.M. Prats, J.I Leon, J.A. Sanchez, J.M. Carrasco, E. Galvan and L.G. Franquelo, "Modeling Strategy for Back-to-Back Three-Level Converters Applied to High-Power Wind Turbines," *IEEE Transactions on Industrial Electronics*, vol 53, no 5, 1483-91, Oct. 2006

[86] P. Kundur, "Power system stability and control," McGraw Hill professional, 1994

[87] "IEEE Guide for planning DC links terminating at AC location having low short-circuit capacities," IEEE Std 1204-1997, 1997

[88] M. Chinchilla, S. Arnaltes and J.C. Burgos, "Control of Permanent-Magnet Generators Applied to Variable-Speed Wind-Energy Systems Connected to the Grid," *IEEE Transactions on Energy Conversion*, v 21, n 1, 130-5, March 2006

[89] "Grid Code, High and extra high voltage," E.ON Netz Gmbh, April 2006

[90] F. Blaabjerg, R. Teodorescu, M. Liserre, and A.V. Timbus, "Overview of Control and Grid Synchronization for Distributed Power Generation Systems," *IEEE Transactions on industrial electronics*, vol. 53, No. 5, October 2006

[91] A.V. Timbus, R. Teodorescu, F. Blaabjerg and M. Liserre "Synchronization Methods for Three Phase Distributed Power Generation Systems. An Overview and Evaluation," *IEEE 36th Power Electronic Specialists Conference*, 2005

[92] A. Milicua, S. Piasecki, M. Bobrowska, K. Rafa, and G. Abad, "Coordinated control for grid connected power electronic converters under the presence of voltage dips and harmonics," in 2009 13th European Conference on Power Electronics and Applications, EPE '09, September 8, 2009 - September 10, 2009, Barcelona, Spain, 2009.

[93] S. Hong-Seok and N. Kwanghee, "Dual current control scheme for PWM converter under unbalanced input voltage conditions," IEEE Transactions on Industrial Electronics, vol. 46, pp. 953-9, 1999.

[94] L.G.B. Rolim, R.D. da Costa, and M. Aredes, "Analysis and Software Implementation of a Robust Synchronizing PLL Circuit Based on the pq Theory, "EEE Transactions on Industrial Electronics, vol 53, No 6, 1919-26, Dec. 2006

[95] S. V. Araújo, A. Engler, B. Sahan and F. L. M. Antunes, "LCL Filter design for grid-connected NPC inverters in offshore wind turbines," International Conference on Power Electronics, Daegu, Korea, 2007

[96] P. Chandrasekhar and S. Rama Reddy, "Design of LCL Resonant Converter for Electrolyser, " International Journal of Electronic Engineering Research, vol 2, No 3, pp.435-444, 2010

[97] I.Martinez de Alegria, J.L.Villate, J. Andreu, I.Gabiola and P.Ibañez, "Grid connection of doubly feed induction generator wind turbines: a survey," EWEC, 2004.

[98] I. M. de Alegríaa, J. Andreua, J. L. Martína, P. Ibañezb, J. L. Villateb and H.Camblongc, "Connection requirements for wind farms: A survey on technical requirements and regulation," Elsevier, Renewable and Sustainable Energy Reviews, vol 11, no 8, pp. 1858-1872, 2007

[99] "Voltage and Transient Stability Support by Wind Farms Complying With the E.ON Netz Grid Code," IEEE transactions on power systems, vol. 22, No. 4, November 2007 1647

[100] "Requisitos de respuesta frente a huecos de tensión de las instalaciones eólicas," REE Procedimiento de operación 12.3, May, 2007

[101] "Procedimiento de verificación, validación y certificación de los requisitos del PO 12.3 sobre la respuesta de las instalaciones eólicas ante huecos de tensión," v.3, REE, 2007

[102] Stephen J. Chapman, "Máquinas elécticas", Mc Graw Hill, 4rd edition, 2004, ISBN 970-10-4947-0.

[103] R. L. García, "Desarrollo y validación de modelos de transformadores monofásicos y trifásicos con saturación, para el análisis de armónicos en sistemas de potencia," Ph. D. Thesis, Universidad politécnica de Cataluña, 2000

[104] A. Pigazo," Método de control de filtros activos de potencia paralelo tolerante a perturbaciones de la tensión de red", Ph. D. Thesis, Universidad de Cantabria, 2004

[105] Monografias. http://www.monografias.com/trabajos21/armonicos/armonicos.shtml

[106] J. A. González, "Compensación de potencia Reactiva en Sistemas Contaminados con Armónicos," Msc Tesis Universidad Central de las Villas, 1998.

[107] "Información Técnica, Generalidades sobre Armónicos," Schneider Electric, 1998.

[108] W.Breuer, N.Christl, "Grid Access Solutions Interconnecting Large Bulk Power On- / Offshore Wind Park Installations to the Power Grid," GWREF, 2006.

[109] J. Plotkin, U. Schaefer, R. E. Hanitsch, "Resonance in the AC Connected Offshore Wind Farms," WECS, 2008.

[110] P. Brogan, "The stability of multiple, high power, active front end voltage sourced converters when connected to wind farm collector systems," Siemens wind power, 2011.

[111] M. Lamich, "Filtros Activos de Potencia," Universitat Polotecnica de catalunya.

[112] L. Yao, "Experience on Technical Solutions for Grid Integration of Offshore Wind farms," DTI Conference Centre, London 2007.

[113] L. Van Der Sluis, "Transients in power systems," John Wiley & Sons Ltd, 2001

[114] ABB Power T&D inc, "Electrical transmission and distribution reference book," fifth edition, 1997

[115] ABB, "Live Tank Circuit Breakers, Application Guide," 2009

[116] R.W. Alexander, D. Dufournet, "Transient recovery voltage (TRV) for high voltage circuit breakers" tutorial, PES

[117] R.G. Garzon, "High voltage circuit breakers, designed and applications," Second Edition, Marcel Dekker, 2002

[118] S. G. Johansson, L. Liljestrand, F. Krogh, J. Karlstrand, and J. Hanson, "AC cable solutions for offshore Wind energy," ABB

[119] C.A. Plet, M. Graovac, T.C. Green and R. Iravani, "Fault response of Grid connected inverter dominated networks," Power and Energy Society General Meeting, Minneapolis, 2010

[120] L. H. Kocewiak, J. Hjerrild and C. Leth Bak, "Wind farm structures inpact on harmonic emission and grid interaction," European Wind Energy Conference, Warsaw, 2010

[121] A. Mazur, I. Kerszenbaum, and J. Frank, "Maximum insulation stresses under transient voltages in the HV barrel-type winding of distribution and power transformers," IEEE Transactions on Industry Applications, vol. 24, pp. 427-33, 1988.

[122] W. Sweet, "Danish wind turbines take unfortunate turn," IEEE Spectrum, vol. 41, pp. 30, 34, 2004.

[123] T. Abdulahovic, "Analysis of High-Frequency Electrical Transients in Oshore Wind Parks," Licentiate of engineering, Chalmers university of technology, Göteborg, Sweden, 2009.

[124] CIGRE working group A2-A3-B3.21, "Electrical Environment of Transformers; Impact of fast transients," ELECTRA, 2005.

[125] R.G. Farner, A.L. Schwalb and E. Katz, "Navajo proyect report on subsynchronous resonance: analysis and solution," *IEEE Transactions on Power Apparatus and Systems*, vol 96, , No 1, pp. 1226-1232, 1997

[126] F. M. Faria da Silva, "Study of high voltage AC underground cables," Ph. D. Thesis, Aalborg University, 2011

[127] "Electrical Transmission and Distribution Reference Book," Westinghouse Electric Corporation; 4th edition, 1964

[128] I.A. Erimez and A.M. Foss, "static synchronous compensator (STATCOM)," Working group 14.19, CIGRE, 2000

[129] "Specifications for Connecting Wind Farms to the Transmission Network," ELTRA Transmission system Planning, April 2000

[130] "Wind Farm Power Station Grid Code Provisions," ESB National Grid, July, 2004

[131] "National Electricity Rules Version 24," Australian energy market comisión, January, 2009

[132] "The grid code," Revision 32, National Grid Electricity Transmission plc, December, 2008

# Appendix A: Nomenclature

## Symbols

| Symbol | Description | Unit |
|--------|-------------|------|
| $v$ | The velocity | [m/s] |
| $h$ | The height | [m] |
| $\alpha_w$ | Roughness length parameter in the current wind direction | |
| $P$ | The power | [Watt] |
| $\phi(v)$ | Weibull's expression for probability density depending on the wind | |
| $k$ | Shape parameter. | |
| $c$ | Scale factor | |
| $\overline{v}$ | The average velocity. | [m/s] |
| $A_r$ | The rotor swept area | [m 2] |
| $r$ | The radius of the rotor | [m] |
| $\rho$ | Density of dry air (measured at average atmospheric pressure at sea level at 15° C) | [kg/m 3] |
| M | The air mass flow | [kg/s] |
| $Cp$ | The power coefficient | |
| R | Resistance | [Ω] |
| V | Voltage | [V] |
| I | Current | [A] |
| $A_C$ | the conductor area of the cable | [m²] |
| $l$ | the length of the cable | [m] |
| $\sigma$ | the conductivity of the cable | [S/m] |
| $\mu_o$ | the magnetic constant or the permeability of the free space | [N/A²] |
| C | the capacity of the cable | [F] |
| $Qe$ | the electric charge stored in a conductor | [Culomb] |
| G | conductance | [Siemens] |
| $\gamma$ | wave propagation constant | |
| $\alpha_\gamma$ | the real part of the propagation constant which represents the attenuation | [Np/m] |
| $\beta_\gamma$ | the imaginary part of the propagation constant which represents phase velocity | [rad/m] |
| $v_e$ | the propagation speed of the traveling wave | [m/s] |
| $f_e$ | the frequency of the analyzed transient phenomenon | [Hz] |
| $\lambda$ | The traveling wave length | [m] |
| Z | the total series impedance of the circuit | [Ω] |
| Y | the total shunt admittance of the circuit | [Ω-1] |
| $Z_c$ | The characteristic impedance | [Ω] |
| $\omega$ | the angular speed | |

| $f$ | The frequency | [Hz] |
|---|---|---|
| $\tau$ | Travelling wave's time delay | [s] |
| $\rho_c$ | the resistivity of the material | [$\Omega$*m] |
| $D$ | The diameter | [m] |
| $\mu$ | the absolute magnetic permeability ($\mu_0\,\mu_r$), | $N/A^2$ |
| $\mu_r$ | the relative magnetic permeability | |
| $A_s$ | The shield's cross section | [$m^2$] |
| $R_s$ | Outer radius of the shield | [m] |
| $r_s$ | Inner radius of the shield | [m] |
| $\varepsilon_r$ | The dielectric constant | |
| $X$ | reactance | [$j\,\Omega$] |
| $Q$ | Reactive power | [VAR] |
| $Ry$ | Rayleigh's probability distribution | |
| $A$ | Avaiavility | [%] |
| $f_{rate}$ | Failure rate | [failure/year] |
| $lf_{years}$ | Life time | [years] |
| $E_{trans}$ | Transmitted energy | [MWh] |
| $C_{platform}$ | Cost of the platform | [M€] |
| $C_{transform}$ | Cost of the transformer | [M€] |
| $C_{comp}$ | Cost of the reactive power compensation | [M€] |
| $C_{invest}$ | Investement cost | [M€] |
| $C_{trans}$ | Transmission cost | [M€] |

## Subscripts

| | |
|---|---|
| *wind* | Wind |
| *HVAC* | High Voltage AC transmission |
| *MVAC* | Medium Voltage AC transmission |
| *cable* | Cable |
| *loss* | Active power losses |
| *active* | Active |
| *reactive* | Reactive |
| *C* | Capacitor |
| *L* | Inductor |
| *comp* | Compensation |
| *max* | Maximum |
| *min* | Minimum |
| *trafo* | Transformer |
| *repair* | repair |
| *avg* | Average |
| *sc* | Short circuit |
| *pcc* | Point off common coupling |
| *pc* | Collector point |
| *ac* | Altern current |
| *t* | Total value |
| *n* | Nominal |
| *in* | Input |
| *out* | Output |
| *loss_avg* | Average active power losses |
| *life* | Usefull life time |
| *rect* | Rectifier |
| *filter* | Filter |
| *bus* | Bus |
| *chopper* | Chopper |
| *d* | Direct |
| *q* | Quadrature |
| *shock* | Shock |
| *dip* | Dip |
| *res* | Residual |
| *TOL* | Tolerance |
| *prim* | Primary |
| *sec* | Secondary |
| *mag* | Marnetization |
| *windfarm* | Windfarm |
| *on* | Onshore |
| *off* | Offshore |
| *resonance* | Resonance |

## Superscripts

| * | Reference |
|---|-----------|
| + | Positive sequence |
| - | Negative sequence |

## Abbreviations

| AC | Altern current |
|----|----------------|
| HVAC | High Voltage AC transmission |
| MVAC | Medium Voltage AC transmission |
| DC | Direct current |
| HVDC | High voltage direct current |
| EMTP | Electromagnetic transients program |
| LCC | Line commutated converters |
| VSC | Voltage source converter |
| MTTR | Mean Time To Repair |
| O & M | Operating and maintenance |
| XLPE | Cross Linked Polyethylene |
| SO | System operator |
| LVRT | Low voltage ride through |
| PCC | Point of common coupling |
| PC | Collector point |
| SCIG | Squirrel cage induction generator |
| WRIG | Wound rotor induction generator |
| PMSG | Permanent magnet synchronous generator |
| WRSG | Wound rotor synchronous generator |
| DFIG | Double fed induction generators |
| PWM | Pulse width modulation |
| THD | Total harmonic distortion |
| PF | Power factor |
| IEGT | Injection Enhanced Gate Transistor |
| PLL | Phase lock loop |
| PI | Proporcional integral |
| SSM | Sequence separation method |
| DSC | Delayed signal cancellation |
| NPC | Neutral point clamped |
| REE | Red Eléctrica Española |
| PVVC | Procedure to verification, validation and certification |
| STATCOM | Static synchronous compensator |
| SVC | Static Var Compensator |

# Appendix B: Power Factor Requirements at the Point of Common Coupling

International grid codes demand to wind farms the control of the power factor at the PCC. Moreover, some of them have as a requirement the control of this power factor depending on the voltage of the PCC.

For example, ELTRA (Denmark), ESB (Irish) and AMC (Australia) grid codes requires a minimum power factor at the PCC independently of the voltage [129], [130], [131] Figure B.1.

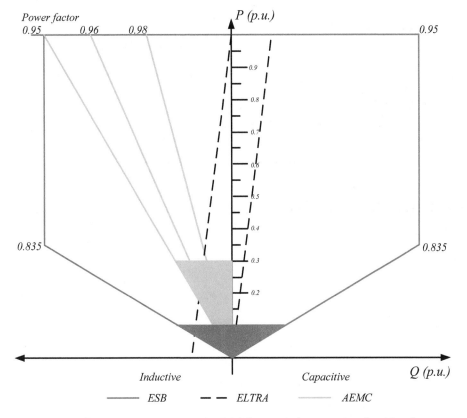

Figure B.1 Power factor requirements at the PCC for several international grid codes.

However, other grid codes: like E.ON or UK national grid, have limited the power factor depending on the voltage in order to contribute to the voltage regulation of the node [132], [89].

As an example, E.ON establishes for normal operation, the power factor boundaries depicted in Figure B.2.

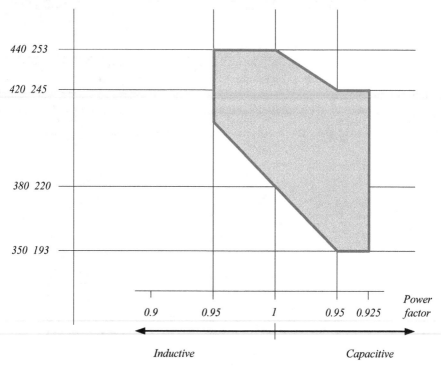

Figure B.2 Power factor requirements in the PCC depending on the voltage for E.ON.

# Appendix C: REE Grid Code Requirements for Voltage Dips

The grid code requirements for voltage dips in Spain are described at the operation procedure P.O. 12.3 (requirements for voltage dips in electrical wind installations) [100] developed by REE (Red Eléctrica de España).

Summarizing the LVRT characteristics required by REE, these can be divided into two fault groups with different requirements (balanced and unbalanced faults) and two different groups of voltage dips:

- Mono phase faults, three-phase faults and two phase-to-ground.
- Two phase ungrounded faults.

In this way, REE grid code specifies that the wind farms (and all their equipments) must not be disconnected, if a three-phase, two phase-to-ground or one-phase faults with characteristics inside the voltage/time curve depicted in Figure C.1 occurs. After fault-clearing the time necessary to recover nominal values depends on the percentage of the wind generation penetration related to the short circuit power

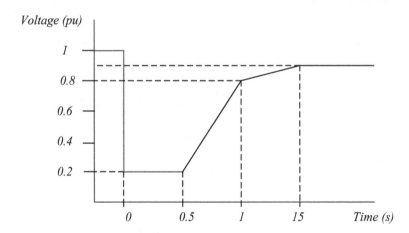

Figure C.1 Voltage / time curve admitted at the PCC for three-phase, two phase-to-ground and one-phase faults

In the case of phase-to-phase short-circuits (two phase ungrounded faults), the maximum voltage drop is 0.6 pu, instead of 0.2 pu. Thus, the wind turbines must not be disconnected if a two phase ungrounded fault with characteristics inside the voltage/time curve depicted in Figure C.2 occurs.

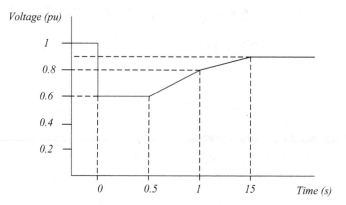

Figure C.2 Voltage / time curve admitted at the PCC for two-phase ungrounded faults

Wind farm also must provide reactive current to the grid at the PCC during the fault and later in the voltage recovery period. In any case, this current must be located in the shaded area in Figure C.3, within 150 ms after the beginning of the fault or after the clearance of the fault. Thus, the wind farm must generate reactive current with voltages below to 0.85 pu, and it must not consume reactive power between 0.85 pu and the rated voltage.

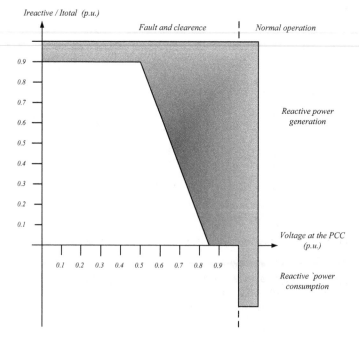

Figure C.3 Reactive current / total current requirement depending on the voltage at the PCC.

## Requirements for balanced three-phase faults

Wind farms will not absorb reactive power during either balanced three-phase faults, or the voltage recovery period after the clearance of the fault. However, reactive power absorptions are allowed during a period of 150 ms after the beginning of the fault, and also 150 ms after the clearance of the fault, with two constrains:

• The net reactive power consumption of the wind farm during the 150 ms interval after the beginning of the fault, in 20 ms cycles, must not exceed 60% of its rated power.

• The net reactive energy consumption of the wind farm after the clearance of the fault must not exceed 60% of its rated power, and the reactive current, in 20 ms cycles, must not exceed 1.5 times the rated current.

With regards to the active power, the wind farm at the PCC must not absorb active power during the fault or the voltage recovery period after the clearance of the fault. However, absorption of active power is accepted for 150 ms after the beginning of the fault and also 150 ms after the clearance of the fault. During the rest of the fault, the active power consumptions are additionally allowed, but have to be less than the 10% of the wind farm rated power.

## Requirements for unbalanced two-phase and single-phase faults

Wind farms will not absorb reactive power in the PCC during either unbalanced two-phase and single-phase faults, or the voltage recovery period after the clearance of the fault. Nonetheless, reactive power absorptions are allowed during a period of 150 ms after the beginning of the fault, and also 150 ms after the clearance of the fault, with two constrains:

• The net reactive power consumption of the wind farm, during the 150 ms interval after the beginning of the fault, will not exceed the 40% of its rated power during a period of 100 ms

• The net reactive power consumption of the wind farm after the fault clearance, in 20 ms cycles, will not exceed the 40% of its rated power.

Additionally, transitory consumption is admitted during the rest of the fault with two constraints:

• The net active consumption must not exceed the 45% of the equivalent rated active energy of the wind farm during a period of 100 ms.

• The consumption of active power, in cycles of 20 ms, must not exceed the 30% of its rated active power.

# Appendix D: Clarke and Park Transforms

The $dq$ transformation is a transformation of coordinates from a three-dimensional stationary coordinate system to the $dq$ two-dimensional rotating coordinate system. This transformation is made in two steps:

- **Clarke Transformation,** a transformation from the three- dimensional stationary coordinate system to a two- dimensional, $\alpha\beta$ stationary coordinate system.
- **Parke Transformation,** a transformation from the $\alpha\beta$ stationary coordinate system to the $dq$ two- dimensional rotating coordinate system.

**Clarke Transformation, from $abc$ to $\alpha\beta$**

A representation of a vector in a three-dimensional space is accomplished through the product of a transpose three-dimensional vector (base) of coordinate units and a vector representation of the vector, whose elements are corresponding projections on each coordinate axis, normalized by their unit values.

If the vector has not any component (the projection on one of the three axes is zero) in one of the coordinate axis, it is possible to transform this vector into an equivalent vector in a two-dimensional space without losing information.

Therefore, if a three-phase space vector in a three-dimensional space has not any component in one of the coordinate axes, the space vector can be transformed into a two-phase space vector in a two dimensional space, in order to simplify the work with them.

Furthermore, if the phase $a$ is arbitrarily chosen to coincide with $a$, the new axes of the two-phase coordinate system, the transformations is even more simple. It can be made applying the equation (150), Figure D.1.

$$X_{\alpha\beta} = X_{abc}T = X_{abc}\frac{2}{3}\begin{bmatrix} 1 & 0 \\ -\frac{1}{2} & \frac{\sqrt{3}}{2} \\ -\frac{1}{2} & \frac{\sqrt{3}}{2} \end{bmatrix} \tag{150}$$

$$X_{abc} = \begin{bmatrix} a_u & b_u & c_u \end{bmatrix} \tag{151}$$

$$X_{\alpha\beta} = \begin{bmatrix} \alpha_u & \beta_u \end{bmatrix} \tag{152}$$

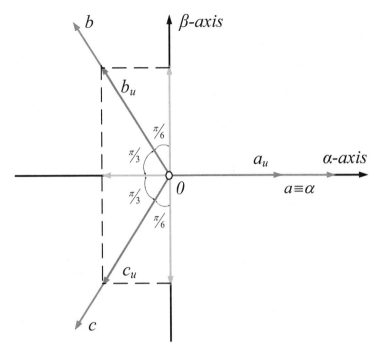

Figure D.1 Vector representation of the *abc* to *αβ* transform.

Without assuming that the three-phase space vector in the three-dimensional space has not any projection on one of the three axes ($0_u{\neq}0$), the three-phase space vector representation transforms to *αβ* vector representation through the transformation matrix defined as:

$$X_{\alpha\beta0} = X_{abc}\,\frac{2}{3}\begin{bmatrix} 1 & 0 & \frac{1}{2} \\ -\frac{1}{2} & \frac{\sqrt{3}}{2} & \frac{1}{2} \\ -\frac{1}{2} & \frac{\sqrt{3}}{2} & \frac{1}{2} \end{bmatrix} \tag{153}$$

$$X_{\alpha\beta0} = \begin{bmatrix} \alpha_u & \beta_u & 0_u \end{bmatrix} \tag{154}$$

**Parke Transformation, *αβ* to *dq***

For the last step, the *αβ* stationary coordinate system is transformed to the *dq* two-phase rotating coordinate system, equation   (155), Figure D.2. This last transformation simplifies the work with rotational three-phase space vectors, due to the fact that if the quadrature axes and the space vector ($X_{\alpha\beta}$) are rotating at the same speed *ωt*, from the rotating point of view of the quadrature axes, the space vector is stationary.

$$X_{dq0} = X_{\alpha\beta0} \begin{bmatrix} \cos\theta & -\sin\theta & 0 \\ \sin\theta & \cos\theta & 0 \\ 0 & 0 & 1 \end{bmatrix} \quad (155)$$

$$X_{dq0} = \begin{bmatrix} d_u & q_u & 0_u \end{bmatrix} \quad (156)$$

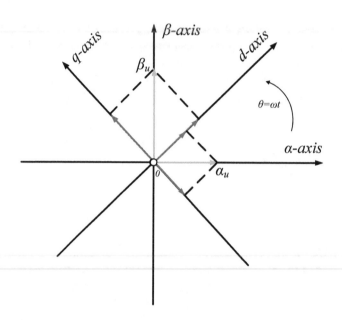

Figure D.2 Vector representation of the $\alpha\beta$ to $dq$ transform.

For the direct transformation, a three-phase vector representation transforms to $dq$ vector representation through the transformation matrix $T$, defined as:

$$X_{dq0} = X_{abc}T = X_{abc}\frac{2}{3}\begin{bmatrix} \cos\theta & \sin\theta & \frac{1}{2} \\ \cos\left(\theta - \frac{2\pi}{3}\right) & \sin\left(\theta - \frac{2\pi}{3}\right) & \frac{1}{2} \\ \cos\left(\theta + \frac{2\pi}{3}\right) & \sin\left(\theta + \frac{2\pi}{3}\right) & \frac{1}{2} \end{bmatrix} \quad (157)$$

And the inverse transformation is:

$$X_{abc} = X_{dq0}T' = X_{dq0} \begin{bmatrix} \cos\theta & \cos\left(\theta - \frac{2\pi}{3}\right) & \cos\left(\theta + \frac{2\pi}{3}\right) \\ \sin\theta & \sin\left(\theta - \frac{2\pi}{3}\right) & \sin\left(\theta + \frac{2\pi}{3}\right) \\ 1 & 1 & 1 \end{bmatrix} \tag{158}$$

# Appendix E: Resonant Passive Filters

The resonant passive filters are constituted by a capacitor, a inductor and a resistor, there are basically electrical branches (RLC branches).

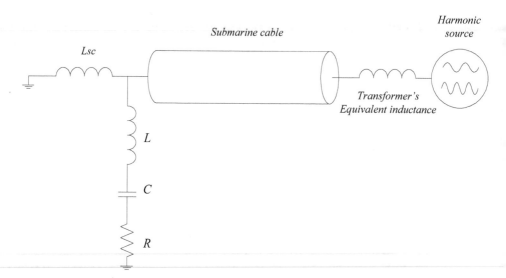

Figure E.1 The transmission system with a RLC filter.

The main characteristic of a RLC branch is the delay of 180° between the voltage dropped in the inductive impedance and the voltage dropped in the capacitive impedance. As a result, in a specific frequency, the electric branch only presents the resistive impedance.

The resonant passive filters are based on this characteristic to filter a specific harmonic. The RLC branch is tuned to presents only the resistive part at the frequency where is located the harmonic. So, at this specific frequency, the RLC branch presents very low impedance and absorbs the harmonic current.

In this way, the harmonic current generated in a device or devices is deviated to the filter instead to flow to the distribution grid or power source. More specifically, the generated harmonic current goes to the grid and to the RLC filter, where the harmonic current is divided depending on kirchoff's law, in the inverse proportion of the impedances. Thus, the filter absorbs more or less of the harmonic current depending on the system impedance and the impedance of the filter at this specific frequency, Figure E.2.

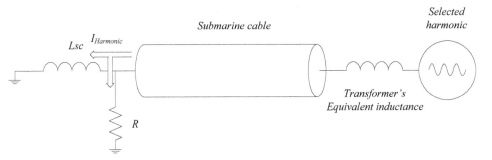

*Submarine cable*

*Selected harmonic*

$L_{sc}$  $I_{Harmonic}$

*Transformer's Equivalent inductance*

$R$

Figure E.2 Transmission system with a RLC filter at the resonance frequency of the filter.

The harmonic current flows through the part of system between the source and the RLC filter. The harmonic current, generates harmonic voltages between the source and the filter, and this harmonic provokes the disturbances described in section 6.2, but the objective of this filter is the reduction of the harmonics waters down the RLC filter. For instance, the reduction of the harmonic levels at the PCC.

Highlight that to achieve a reduction bigger than the 50% of any harmonic current with this kind of filters (a RLC branch in parallel with the circuit), the impedance of the filter at the selected frequency have to be less than the impedance presented by the circuit at this frequency.

## RLC impedance depending on the frequency

To carry out the evaluation of the impedance, current and voltage through the RLC branch depending on the frequency, there is considered the generic RLC filter depicted in Figure E.3

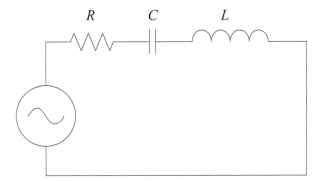

$R$  $C$  $L$

Figure E.3 Generic RLC circuit.

Looking at Figure E.4, the diagram of the RLC branch voltages shows a voltage drop in phase with the current in the resistance, a voltage drop delayed 90° with the current in the capacitor and a voltage drop (delayed -90°) 90° forward the current in the inductor.

The voltage dropped in the capacitor has a difference of 180° with regard to the voltage dropped in the inductor, so, the both voltages are counteracting each other. Consequently, the reactive voltage vector of the RLC branch is the subtraction of these two vectors $V_{XL} - V_{XC}$ (or $V_{XC} - V_{XL}$).

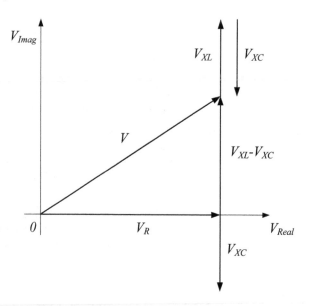

Figure E.4 Diagram of the voltages of the RLC branch.

Therefore, the total impedance of the RLC branch in complex form can be calculated via equation (159).

$$Z = R + \left(jX_L - jX_C\right) \tag{159}$$

**Resonance**

A circuit is (or goes to) in resonance when the applied voltage and the current through the circuit are in phase. Thus, it is possible to conclude that in resonance, the total impedance of the circuit is equal to its resistive part, i.e. the reactive impedance of the capacitor and the reactive impedance of the inductor are the same, but delayed 180°.

In short, in resonance, the reactive part of the circuit impedance has to be zero. This occurs for a specific frequency value and this frequency is the so called resonance frequency. In the RLC circuit, there are two independent parameters: $L$ and $C$. Thus, there are infinite combinations to obtain a resonance in any specific frequency.

At Figure E.5, the impedance of each component of the electric branch (inductive, capacitive and resistive) depending on the frequency is represented.

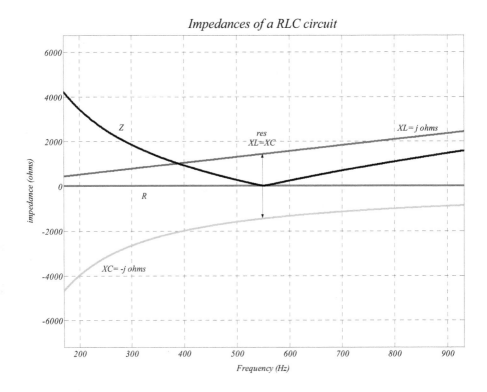

Figure E.5 Impedance of each component of the RLC branch depending on the frequency.

As can be seen in Figure E.5, the resistive part of the total impedance ($R$) is constant, i.e. it is not frequency dependent. However, the inductive part of the total impedance ($ZL$) grows up linearly with the frequency and the capacitive part ($ZC$) grows up exponentially with the frequency from "minus infinite" (for zero Hz) to zero (for an infinite value of frequency).

So, the total impedance of the branch ($Z$) decreases to the value of the resistive component (at the resonance frequency) and then grows up again. Thus, the resistance is the impedance which limits the current when the circuit is in resonance.

To characterize in more detail this relation between the resistive component and the resonance, the admittance ($Y = 1/Z$) of the RLC branch depending on the frequency for different R values is depicted in Figure E.6.

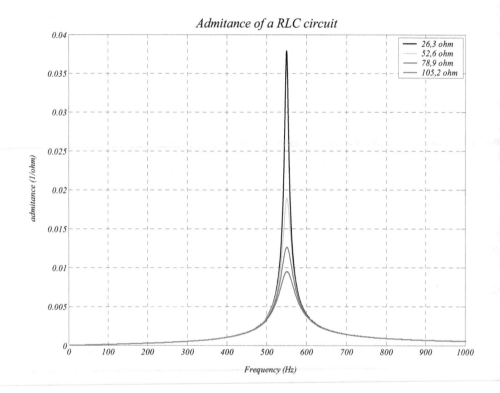

Figure E.6 The admittance of the RLC branch depending on the frequency for several values of the resistive part.

In Figure E.6, it is possible to observe that the relation between the amplitude of the resonance and the amplitude of the resistance is inversely proportional, i.e. if the smaller is the resistive part, the RLC branch can accept more current at this frequency.

### Quality factor

Is called the quality coefficient or quality factor to the product of the pulsation and the relation between the maximum stored energy and the average dissipated power. This quality factor is labeled as $Q$ and its expression is as follows:

$$Q = \frac{fo}{\Delta f} = \frac{1}{Rf} \cdot \sqrt{\frac{Lf}{Cf}} \tag{160}$$

Where: $\Delta f$ is the band width.

In the same way, $Q$ is defined as the relation between the voltage drop in the coil (or the capacitor) and the resistor. Usually, this factor has values above to 10. It is possible to see in Figure E.6, how the quality of a circuit is bigger as smaller is the resistor.

## Criterions for the RLC passive filter design

To characterize adequately the resonant passive filters, there are must be taken into consideration the following aspects:

1. Power losses at fundamental frequency.
2.-Reactive power generated at fundamental frequency.
3.-The quality factor ($Q$) and the band width ($\Delta f$).
4.-Resonance frequency of the filter.
5.-The maximum harmonic voltage and harmonic current capable to support the filter, i.e. the nominal voltage and current for different components of the filter.

Notice that all of these aspects are inter related.

## Active power losses mitigation

As is mentioned, resonant passive power filters have active power losses at fundamental frequency. These losses are continuous and can be significantly high. One option to reduce significantly these losses is placing an inductance in parallel with the resistor, Figure E.7.

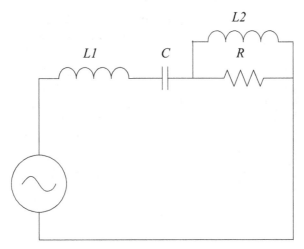

Figure E.7 Generic (R+L)LC circuit.

This inductance (L2) has to be dimensioned to present a small impedance at fundamental frequency (R>>XL2) and to present a high impedance (XL2>>R) at resonance frequency, Figure E.8.

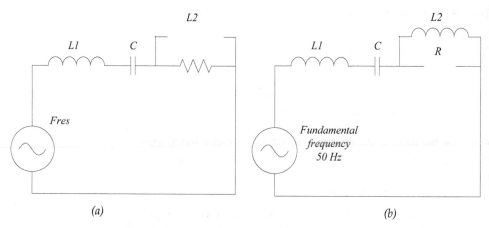

(a)                                                              (b)

Figure E.8 Equivalent (R+L)LC circuit approximation depending on the frequency, (a) At resonance frequency and (b) at fundamental frequency.

The equivalent circuits represented in Figure E.8, shows the philosophy of this method to reduce the active power losses of the filter. Nevertheless, this representation is an approximation. Depending on the resonant frequency, this approximation will be more accurate or less.

If the resonant frequency of the branch is significantly high in comparison with the fundamental frequency ($f_{res} >> f_{fundamental}$), the difference between the impedance presented by the inductance (L2) at the fundamental frequency and at the resonant frequency is also significantly big. Thus, in this cases is possible to adjust the inductance L2 to be small at fundamental frequency and big at resonant frequency.

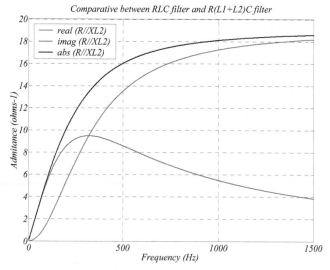

Figure E.9 Evolution of the (R+L2) impedance depending on the frequency.

The evolution depending on the frequency for the equivalent impedance presented by the inductance and the resistor (R+L2) is illustrated in Figure E.9.

Looking at Figure E.9, at low frequencies, the absolute impedance is clearly inductive (high influence of the inductance) which reduces the active power losses. However, at high frequencies, the absolute impedance is almost the same of the real value (high influence of the resistor), allowing the resistor limiting the maximum current through the branch at resonance frequency.

*The new resonance frequency*

As can be seen in Figure E.9, at high frequencies, the absolute impedance is almost the same of the real value, but there is still an inductive part. Therefore, this inductive component varies (reducing) the resonant frequency of the branch, Figure E.10.

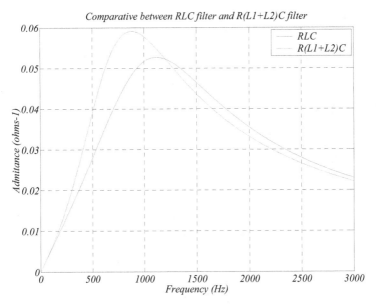

Figure E.10 Evolution of the admittance depending on the frequency. (Red) for the RLC branch and (b) for the (R+L)LC branch.

The new resonance frequency can be estimated in the same way as did for the RLC branch, due to the fact that the resonance occurs at the frequency where the inductive impedance is exactly the same of the capacitive impedance.

The only difference of this case comparing with the previous case is that the inductive impedance of the branch is given by the sum of the L1 inductance and the inductive part of the (R+L2) impedance, Figure E.11.

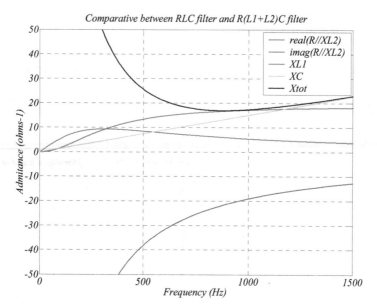

Figure E.11 Impedance of the (R+L)LC branch depending on the frequency and for each component.

Therefore, varying (reducing) the impedance of (L1) it is possible to keep the resonance frequency of the (R+L)LC at the same frequency of the RLC branch. The new value for L1 is given by the subtraction of the inductive component of the (R+L2) impedance at the resonance frequency to the impedance of the capacitor, equations (161) - (163).

$$Z_{equi} = \frac{R + jX_{L2}}{jRX_{L2}} + \tag{161}$$

$$jX_{L1}(f_{res}) = jX_C(f_{res}) - imag(Z_{equi}(f_{res})) \tag{162}$$

$$L1 = \frac{X_{L1}}{2\pi f_{res}} \tag{163}$$

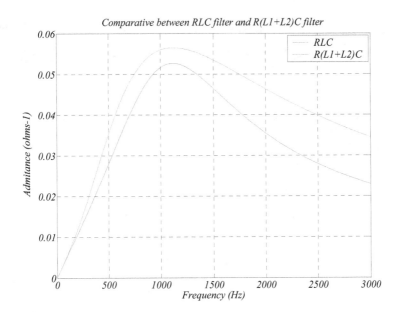

Figure E.12 Evolution of the admittance depending on the frequency. (Red) for the RLC branch and (b) for the equivalent (R+L)LC branch.

As can be seen from Figure E.12, the (R+L)LC branch has a different admittance evolution in comparison with the evolution of the RLC branch, the (R+L)LC branch has bigger band width.

Therefore, as is mentioned, this method is more appropriate for filters tuned for high resonance frequencies, the higher is the resonance frequency in comparison with the fundamental, the more similar is the evolution of the impedance of the (R+L)LC branch for the RLC branch.

# Appendix F: Comparison and Validation of the Equivalent Feeder

The objective of the present appendix is to validate the simplification carried out to define the considered scenario for the problem assessment in chapter 7, section 7.2. Thus, in order to validate the simplification, a comparative between the equivalent wind turbine and the full feeder (feeder composed by six wind turbine models explained in chapter 5, section 5.2.2) is performed. In this way, it is possible to compare the behavior of these two systems and corroborate that the equivalent wind turbine exhibits a reasonably similar behavior.

For this purpose, the simulation results of the two scenarios depicted in Figure F.1 upon three-phase (80% of depth) and two-phase (40% of depth) faults are compared.

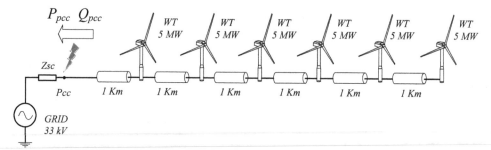

*(a)*

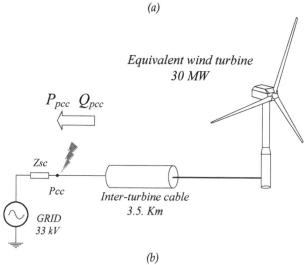

*(b)*

Figure F.1 Considered simulation scenarios to validate the simplification. (a) Full feeder and (b) equivalent wind turbine.

In the first step, the simulation of a three-phase fault at the PCC for both scenarios is carried out.

All the wind turbines are working at full load (90% of the nominal power, Table 5.13) and a power factor of 0.95 inductive at the PCC. The simulation results for this first case are shown in Figure F.2 - Figure F.6.

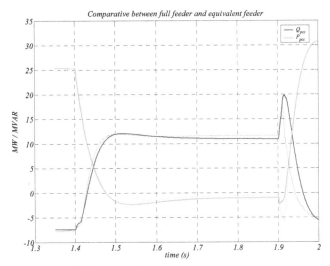

Figure F.2 Comparison of the equivalent wind turbine and the full feeder for a three-phase fault, the evolution of the active and reactive power during the fault at the PCC,

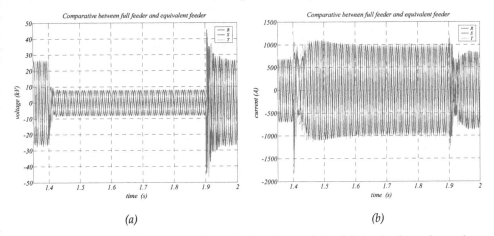

(a)                                        (b)

Figure F.3 Comparison of the equivalent wind turbine and the full feeder for a three-phase fault, the evolution of the voltage (a) and current (b) during the fault and the clearance at the PCC.

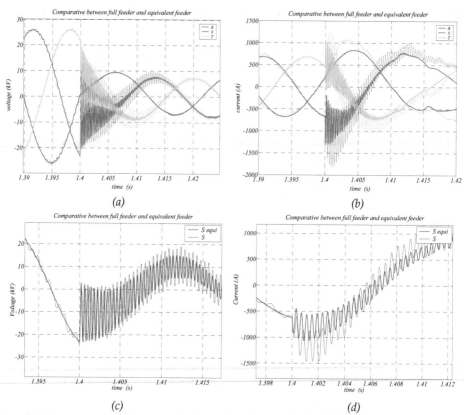

(a)                                                              (b)

(c)                                                              (d)

Figure F.4 Comparison of the equivalent wind turbine and the full feeder for a three-phase fault, the voltage (a) and current (b) at the PCC the same instant that the fault occurs, (c)-(d) at the same instant of the fault mono-phase and more detailed (red, equivalent feeder).

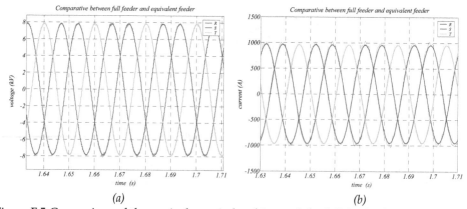

(a)                                                              (b)

Figure F.5 Comparison of the equivalent wind turbine and the full feeder for a three-phase fault, the voltage (a) and current (b) at the PCC during the maintenance of the fault.

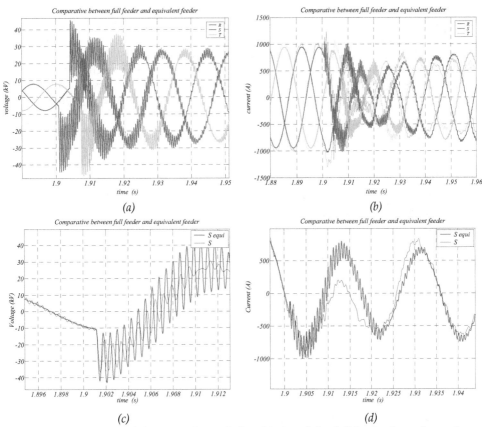

*(a)*  *(b)*  *(c)*  *(d)*

Figure F.6 Comparison of the equivalent wind turbine and the full feeder for a three-phase fault, the voltage (a) and current (b) at the PCC when the clearance of the fault occurs. (c)-(d) the voltage and current at the clearance of the fault mono-phase and more detailed, (red, equivalent feeder).

The submarine cable is simplified to generate the same reactive power, the same voltage drop and the same active power loses (equation (144)). Consequently, the results depicted in Figure F.2 are very similar.

Current and voltage peaks during the transient are caused by the energizing and de-energizing of the cable (see section 7.1). The equivalent feeder has less capacitive component and less resistive part but the same short circuit impedance. As a result, the equivalent feeder has smaller inrush current peak at the beginning of the fault (the de-energizing). The less is the capacitive component, the less is the stored energy and as a consequence needs less current for de-energizing.

Due to the fact that needs less inrush current, it is possible to see fewer low frequency oscillations before the system reaches the steady-state in the results of the equivalent feeder.

The clearance of the fault is at zero current and both the equivalent and the full feeder are connected to the grid with the same short circuit inductance, thus, both systems have the same energy stored at the magnetic field of this inductance. Therefore, the inductance transferred the same energy from the magnetic field to the electric field to adapt to the new steady-state. But, the equivalent feeder has less resistive component and less capacitive component (needs less energy to be energized). So, in contrast to the beginning of the fault, at the clearance of the fault the equivalent wind turbine has more low frequency oscillations.

Therefore, the way that is simplified the submarine cable, which varies the cable length (capacitive component of the cable), explains the difference in both transients.

Nevertheless, in steady-state the voltage and the current of the equivalent feeder have less high frequency oscillations, because is modeled with an ideal voltage source.

In the second step, the simulation of the two-phase fault is performed. The results for the equivalent wind turbine and the full feeder for this second case are depicted in Figure F.7 - Figure F.11.

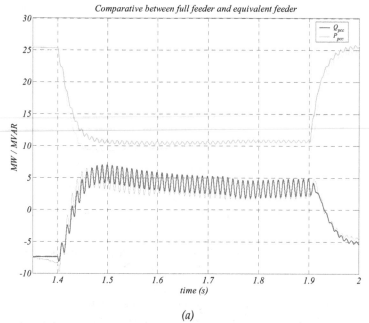

*(a)*

Figure F.7 Comparison of the equivalent wind turbine and the full feeder for a two-phase fault, the evolution of the active and reactive power during the fault at the PCC.

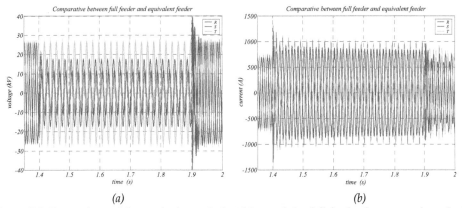

(a) (b)

Figure F.8 Comparison of the equivalent wind turbine and the full feeder for a two-phase fault, the evolution of the voltage (a) and current (b) at the PCC during the fault and the clearance.

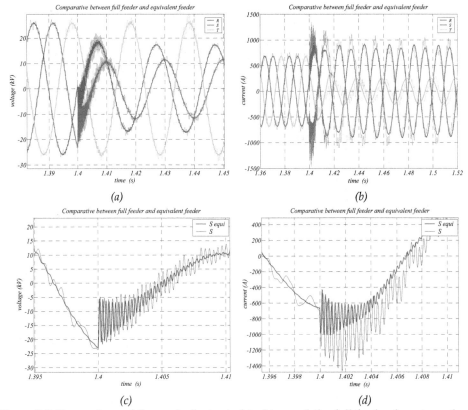

(a) (b)

(c) (d)

Figure F.9 Comparison of the equivalent wind turbine and the full feeder for a two-phase fault. the voltage (a) and current (b) at the PCC the same instant that the fault occurs, (c)-(d) at the same instant of the fault mono-phase and more detailed (red, equivalent feeder).

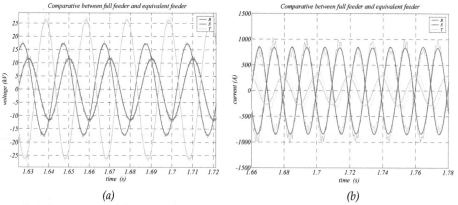

Figure F.10 Comparison of the equivalent wind turbine and the full feeder for a two-phase fault, the voltage (a) and current (b) at the PCC during the maintenance of the fault.

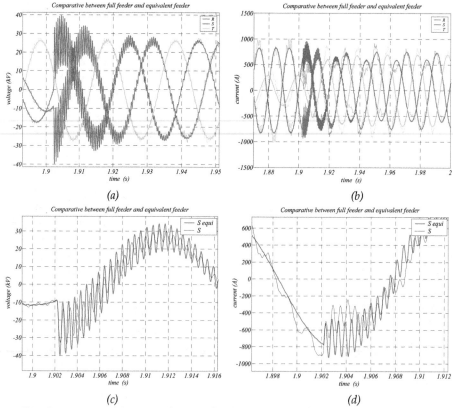

Figure F.11 Comparison of the equivalent wind turbine and the full feeder for a two-phase fault, the voltage (a) and current (b) at the PCC when the clearance of the fault occurs. (c)-(d) the voltage and current at the clearance of the fault mono-phase and more detailed, (red, equivalent feeder).

As is explained before, the transient response exhibits the typical behavior for a fault in a RLC circuit described in section 7.1. With regards to the frequency of the oscillation, the energy exchange between capacitive component of the cable and the short circuit impedance of the grid causes the oscillation. Therefore, this frequency depends on those impedances.

This fact (the dependence of the resonance and the oscillation frequency with the capacitive component / the total length of the inter-turbine grid) strengthens the decision to simplify the wind farm with equivalent wind turbines for each feeder and not only one. Because, using only one equivalent wind turbine for the entire wind farm, with only an equivalent inter-turbine cable (less capacitive component), will change this oscillation frequency.

Obviously, the simplification carried out also changes the oscillation frequency and the de-energizing current peak for the cable, due to the fact that only takes into account a part of the inter turbine cable (equation (144)), not the whole cable. But, is better simplification than only one equivalent wind turbine which takes into account only a very little part of the total capacitive component of the real inter-turbine grid.

Looking to the results obtained in both scenarios, it is possible to see a similar oscillation frequency and a reasonable similar current/voltage peaks. Moreover, roughly the evolution of the voltage and current are similar too. Thus, based on the results depicted in Figure F.2 - Figure F.6 and Figure F.7 - Figure F.11, it is possible to conclude that the equivalent wind turbine is a reasonable accurate simplification.

# Appendix G: Considered STATCOM Model to Validate the Proposed Solution

The objective of the present appendix is to characterize the model and characteristics of the considered STATCOM used in the simulation scenario to validate the proposed electric connection infrastructure (chapter 7, section 7.5.4).

This simulation is oriented mostly to measure the injected reactive power by the offshore installation to the main grid (at the PCC). Therefore, the most important characteristic of the STATCOM is its rated current/power, i.e. the capability of the STATCOM to inject reactive power during voltage dips at the PCC. Due to this fact, the details of the model and control of the STATCOM are not much relevant.

In this way, for the sake of simplicity, there are considered similar features for the STATCOM and for the grid side converters of the wind turbines: Same topology for the converter (a three-level topology with a 3.3kV ($v_{statcom}$) output voltage based on IEGTs, see section 5.2.2.1 ), same control strategy (two proportional-integral gains in the $d$-$q$ frame for each sequence with cross-coupled terms, see section 5.2.2.2 ) and the same parameters to tune the LC-L filter (see section 5.2.2.3).

Thus, the considered STATCOM used in the simulation scenario to validate the proposed electric connection infrastructure is shown in Figure G.1.

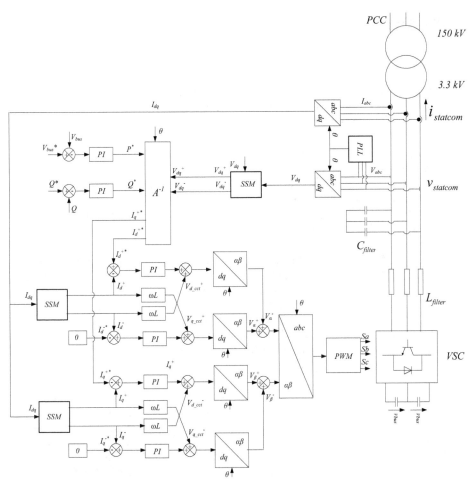

Figure G.1 Main scheme of the considered STATCOM used to validate the proposed electric connection infrastructure.

# Permissions

The contributors of this book come from diverse backgrounds, making this book a truly international effort. This book will bring forth new frontiers with its revolutionizing research information and detailed analysis of the nascent developments around the world.

We would like to thank M. Zubiaga, G. Abad, J. A. Barrena, S. Aurtenetxea and A. Cárcar, for lending their expertise to make the book truly unique. They have played a crucial role in the development of this book. Without their invaluable contribution this book wouldn't have been possible. They have made vital efforts to compile up to date information on the varied aspects of this subject to make this book a valuable addition to the collection of many professionals and students.

This book was conceptualized with the vision of imparting up-to-date information and advanced data in this field. To ensure the same, a matchless editorial board was set up. Every individual on the board went through rigorous rounds of assessment to prove their worth. After which they invested a large part of their time researching and compiling the most relevant data for our readers. Conferences and sessions were held from time to time between the editorial board and the contributing authors to present the data in the most comprehensible form. The editorial team has worked tirelessly to provide valuable and valid information to help people across the globe.

Every chapter published in this book has been scrutinized by our experts. Their significance has been extensively debated. The topics covered herein carry significant findings which will fuel the growth of the discipline. They may even be implemented as practical applications or may be referred to as a beginning point for another development. Chapters in this book are authored by M. Zubiaga, G. Abad, J. A. Barrena, S. Aurtenetxea and A. Cárcar, first published by InTech; hereby published with permission under the Creative Commons Attribution License or equivalent.

The editorial board has been involved in producing this book since its inception. They have spent rigorous hours researching and exploring the diverse topics which have resulted in the successful publishing of this book. They have passed on their knowledge of decades through this book. To expedite this challenging task, the publisher supported the team at every step. A small team of assistant editors was also appointed to further simplify the editing procedure and attain best results for the readers.

Our editorial team has been hand-picked from every corner of the world. Their multi-ethnicity adds dynamic inputs to the discussions which result in innovative outcomes. These outcomes are then further discussed with the researchers and contributors who give their valuable feedback and opinion regarding the same. The feedback is then collaborated with the researches and they are edited in a comprehensive manner to aid the understanding of the subject.

Apart from the editorial board, the designing team has also invested a significant amount of their time in understanding the subject and creating the most relevant covers. They scrutinized every image to scout for the most suitable representation of the subject and create an appropriate cover for the book.

The publishing team has been involved in this book since its early stages. They were actively engaged in every process, be it collecting the data, connecting with the contributors or procuring relevant information. The team has been an ardent support to the editorial, designing and production team. Their endless efforts to recruit the best for this project, has resulted in the accomplishment of this book. They are a veteran in the field of academics and their pool of knowledge is as vast as their experience in printing. Their expertise and guidance has proved useful at every step. Their uncompromising quality standards have made this book an exceptional effort. Their encouragement from time to time has been an inspiration for everyone.

The publisher and the editorial board hope that this book will prove to be a valuable piece of knowledge for researchers, students, practitioners and scholars across the globe.

Printed in the USA
CPSIA information can be obtained
at www.ICGtesting.com
JSHW011434221024
72173JS00004B/799